D1233061

The Economics of Invention

The Economics of Invention

A Study of the Determinants of
Inventive Activity

Geoffrey Wyatt

Senior Lecturer in Economics
Heriot-Watt University

ST. MARTIN'S PRESS   New York

First published in the United States of America in 1986

Printed in Great Britain

ISBN 0-312-23655-7

Library of Congress Cataloging-in-Publication Data

Wyatt, Geoffrey.
  The economics of invention.

  Bibliography: p.
  Includes index.
  1. Inventions—economic aspects. I. Title.
T212.W93   1986   338.4'7608   86-3945
ISBN 0-312-23655-7

to Sinikka

Contents

Preface

There are not many books on the economic determinants of invention, which is rather surprising if one considers the extent to which changes in technology have affected our present standard of living, and the likelihood that they will continue to do so. Perhaps the dearth is due to a feeling that the occurrence of inventions has little to do with economics. I hope that this book will convince some readers that this need not be the case.

Though the first chapter is a rather extensive introduction and summary to the remaining chapters, and can provide the busy reader with a convenient overview, it may be useful at this point to state the general framework within which the various themes are developed. The mode of analysis is 'neoclassical', in that it assumes maximising behaviour on the part of individuals, and market-determined allocations of resources. The economics is also, in the main, positive rather than normative. It may be possible to have an economic analysis of invention which is not based on 'the neoclassical paradigm', and there are plenty of normative issues to engage economists in this area. But the emphasis here is on the determinants, rather than the consequences, of inventive activity, and I believe that positive, neoclassical economics is adequate for the task.

This book is a revised version of the first eight chapters of a thesis which was eventually submitted to the University of York in 1984. The other two chapters of that thesis comprise the essence of a monograph published by the Research Institute of the Finnish Economy (ETLA) under the title *Multifactor productivity change in Finnish and Swedish industries, 1960 to 1980* (ETLA, Helsinki, 1983).

It is my privilege at this point to acknowledge debts to various people who have been helpful to me along the way. At York in the far-off days of the late 1960s John Williamson provided the initial stimulus, and Keith Hartley carried on where he left off, probably for longer than he cares to recall. Tangible support, in the form of research employment, was initially provided by Jack Wiseman, the director of the Institute for Social and Economic Research at York, and more recently by Pentti Vartia, the director of ETLA in

Helsinki where I spent an enjoyable few months in 1982. I am pleased that ETLA is including this book in their series of thesis-type publications, and I am grateful to a number of people there for support in bringing it to publication, in particular to Pekka Ylä-Anttila who heads the industrial economics group, and to Arja Selvinen and Arja Virtanen who drew the numerous diagrams. My thanks also to Jean Roberts, who typed the manuscript in various drafts. I am also grateful to the editorial staff of Wheatsheaf publishers for their expeditious handling of this joint publication.

Geoffrey Wyatt
Edinburgh,
November, 1985

Part I
Theoretical

1 Introduction and Overview

1.1 PRELIMINARIES AND CAVEATS

An invention is an addition to the stock of factual knowledge. It may be that some inventions 'just happen', but most do not; normally they are the outcome of a research process. Inventive activity is a form of research. This book, too, is a product of research, namely research into research. As usual, many of the insights reported here are truly the work of others: most research builds on prior research. To a large extent, therefore, these pages represent a synthesis, a pulling together, of what is known or conjectured about invention from the standpoint of economics. But it is not a comprehensive survey of the economics of invention. Rather, in order to focus on a theme, 'the economics of invention' has been interpreted in a particular way, and the first task of this introduction is to clarify how.

This book is about an aspect of the economics of technological change. But technological change has much wider connotations than the coverage here implies. Invention is construed as anything that adds to the set of known technological possibilities. Actual changes in technology may, however, also derive from a fuller or different utilisation of technological possibilities already in existence. Such changes do not require inventions. They are accordingly not discussed here. This means, of course, that the diffusion of new techniques of production, or the imitation of innovations, are ruled out. Nor is innovation itself, even to the extent that it implies the putting into effect of inventions, discussed more than cursorily.

It should be clear therefore that the present work has little to say about the consequences of invention as such. The focus is on the determinants of invention and inventive activity, on its causes rather than consequences. The upshot of this is that there are, undoubtedly, many aspects of economic life that are affected by invention which find no mention here. Thus, there is no discussion of technological unemployment, of the structure of industry and how it is affected by technological change, of the consequences for economic growth of endogenous technological change and so on.

3

(For a good discussion of many of these issues, the reader should refer to Stoneman (1983).)

An implication of focusing on the determinants of invention rather than its effects is that there is only limited discussion of normative or welfare issues. The bulk of the book is positive in the methodological sense of describing or theorising about what is rather than what ought to be. An exception to this is Chapter 5, on invention market organisation, where it is argued that the peculiar economic characteristics of invention imply the need for considerable regulation. There the desirability of competition between inventors is discussed, but the institutional framework within which inventions are produced, and in particular the assumption that patents can confer ownership on them, is taken as given. There is no extended discussion of the desirability of patents or of possible alternative social arrangements.

There are, of course, many factors that determine how much inventive activity is carried out and in what directions, but the basic supposition is that among these, importantly, are economic factors. This means that, for the most part, the scientific, technological, sociological, political, psychological and cultural determinants of inventive activity are subsumed in an implicit *ceteris paribus* clause. The present analysis, in other words, takes them as given. The realism of this assumption can only be judged with suitable empirical evidence. The empirical analysis of time-series data on numbers of patents for inventions presented in Chapter 6 is consistent with the view that at least a substantial element of inventing takes place in response to economic stimuli. But of course this does not imply that scientific, sociological, etc. factors are unimportant. There is, in the present work, no attempt made to assess the relative importance of all possible determinants of invention.

Thus the focus is on the economic determinants of inventive activity, and they are located specifically at the microeconomic level. It is argued theoretically in Chapter 3, with empirical support in Chapter 7, that the 'level of activity' is a major determinant of inventive activity. Here, this expression refers not to the conjunctural state of the national economy but to the size, measured by output or factor inputs, of the industry or sector to which the invention relates.

As a final caveat on the limited scope of the book, it should be noted that what has come to be known as 'the Schumpeterian hypothesis' is barely touched on. The Schumpeterian hypothesis holds that larger firms are more progressive than their smaller counterparts, in the sense of being more able and willing to employ research inputs. The received empirical wisdom is that neither the smallest nor the largest firms in an industry are the most progressive in this sense. Investigating the validity of Schumpeter's conjecture has been a major preoccupation of economists, since it could have a profound influence on public policy towards big business. If true, it seems to imply a trade-off between static and dynamic efficiency. The topic is touched on in Chapter 3, only to the extent that market organisation has a bearing on the derived demand for invention at the industry level.

It has been mentioned that the central theme of this book is the responsiveness of invention to economic factors. This is developed in a neoclassical framework of supply and demand analysis. There are other approaches which provide their own insights. One such is the comparative institutions approach, according to which allocations of resources are determined either by decentralised markets or hierarchical organisations depending on which structure is the most cost-effective in the presence of behavioural constraints such as bounded rationality and opportunism, (see Williamson, 1975). Another, not unrelated, alternative approach is to eschew the full rationality assumption of the neoclassical framework in favour of 'satisficing' or rule-of-thumb decision-making behaviour. This has the disadvantage of seldom providing clear analytical predictions, though qualitative conclusions can be reached by simulation (cf. Nelson *et al.*, 1976).

Within a neoclassical framework for the determinants of invention, the pivotal element is supply. But it will be seen in Part I that there is little that can be established about the supply of inventions from an *a priori* standpoint. In comparison, economic analysis offers much greater scope on the derived demand for inventions. The basic idea that the derived demand for an invention is proportional to its extent of application is the basis for the attempt in Part II to form an empirical assessment of the responsiveness of invention supply to economic factors.

1.2   SYNOPSIS OF PART I

Part I, comprising Chapters 2–5, is theoretical. Chapter 2 examines the nature of inventive activity and its output, initially taking a taxonomic approach. The conventional classification of research into basic, applied and development has only limited usefulness for an economic analysis of invention. A more important and fundamental distinction is between agenda-reducing and agenda-increasing research, where research output is new knowledge. Building on this distinction, which is due to Fritz Machlup, it is argued that agenda-reducing research can in principle be measured by the reduction in entropy of a probability distribution over possible states of nature, where the probabilities have a subjective degree of belief interpretation. But there is no equivalent way of measuring the output of agenda-increasing research activity. As a consequence, despite the fact that the insights of information theory provide a useful vehicle for characterising research activity, it cannot provide an adequate measure of research output. When invention and innovation are treated together, a satisfactory measure of such output is the rate of cost reduction or, where new products are concerned, the increase in efficiency of satisfying consumer wants.

The basic feature of research, being an exploration of the unknown, provides scope for several related paradigms of the research process. Thus search and sampling models are presented in an attempt to make qualitative statements about the invention supply function. On this basis it is suggested that the research production function can be expected to exhibit sharply diminishing returns for a given state of scientific knowledge.

Research output is not wholly characterised by a mere description of the inventions produced: when they are produced is also important. Thus research output is an example of joint production – what and when. This points to an inherent feature being the time–cost trade-off. The final section of Chapter 2 presents a derivation of the trade-off from the sampling paradigm of research activity. In summary, Chapter 2 represents an attempt to present a coherent view of the research process from an *a priori* standpoint. It goes some way to underpinning the ideas of diminishing returns to the 'research production function' and the convexity of the time-cost trade-off that are often assumed as the starting-point of analysis in this area.

In Chapter 3, the focus is switched from the supply to the demand for invention. The basic determinants of demand are the size of the invention-using industry, the structure of costs and the prices of factors of production, together with the market organisation of that industry, and the appropriability of the returns to invention. These determinants of the derived demand for invention, excluding the influence of factor prices which is deferred to Chapter 4, are discussed in some detail. Previous analysts, beginning with Kenneth Arrow and Harold Demsetz, had focused on the apparently normative question, whether competition or monopoly in the final product market give the greater incentive to invent. Demsetz's conclusion – that the difference between them is due to the output-restricting effect of monopoly – confirms the leading role of industry size. From a positive point of view, however, the question is not whether there are many or few firms in the industry but whether there are barriers to entry. As the theory of contestable markets has shown, freedom and costlessness of entry and exit, together with fungibility of fixed cost, imply that the competitive solution can be an equilibrium even for 'natural monopolies'.

The basic analysis with limit-pricing of the invention, which is familiar in the literature, is presented in section 3.2. As elsewhere, it is assumed throughout Chapter 3 that the inventor has patent protection. When the invention-using industry is a monopoly with entry barriers this gives rise to a bilateral monopoly between the demander and the supplier of the invention. It is shown in 3.3 that it is in the production monopolist's interest to internalise the research, or else to treat the payments to the external inventor as a fixed cost of production.

In 3.4 a full neoclassical treatment of the derived demand for invention, using the fundamental propositions of duality theory, is presented. In 3.5 the degree of appropriability of the returns to invention is made a variable via the duration of patent protection. The question whether, for a given demand curve for the final product, competition or entry-protected monopoly provides the greater incentive to invent is shown to hinge on the relative permanence of the product monopoly and the monopoly implicit in the patent.

The implications for incentives to produce capital-embodied inventions when marginal costs reflect the past history of capital-embodied technologies, as postulated by W. E. G. Salter, are examined in section 3.7. Finally in this chapter, the implications for the demand price of invention when the invention user is a protected monopoly whose goal is sales revenue-maximisation is explored.

None of the variations on the basic analysis overturns the conclusion that the demand for invention is a function of the scale of the industry. This is the most important conclusion to emerge from Chapter 3, and extensive use of it is made in the empirical sections in Part II.

It was seen above that the set of prices of factors of production in the invention-using industry is a determinant of the demand price of the invention. The responsiveness of invention characteristics to factor prices, which is referred to as the induced bias of invention, has been a distinct strand in the history of thought in this area. When factor prices change, there is a tendency to bring in new known techniques of production that imply less intensive use of the factors whose relative prices have increased. If there are also substitution possibilities in the production of inventions which imply different factor intensities in production, then invention displays an induced bias. Empirical evidence on factor-saving bias is presented in section 4.3. This presents a summary of the existing literature on this topic, and belongs naturally in Chapter 4 rather than in Part II as no new empirical results are presented.

The analysis in Chapter 4 assumes that inventions represent a particular variety of changing technology, known as factor augmentation. A model of induced invention is then naturally characterised in terms of a trade-off between attainable rates of factor augmentation. This is known as the 'invention possibility frontier'. some doubts have been expressed in the literature about the shape of this frontier, but it is argued in 4.4 that, by its analogy with a conventional isoquant, it is natural to assume that it has the postulated concave shape. The invention possibility frontier has played an important role in some macroeconomic growth theories as it implies an asymptotically Harrod-neutral equilibrium growth path. It is argued in 4.5 however that in a microeconomic context it tends to induce Hicks-neutral technological change in the long run. The difference between the micro and macro implications of the frontier is due to the fact that factor prices are treated parametrically at the micro level, whereas at the macro level it is factor supplies that are given.

Chapter 4 concludes by observing that in a world in which

technological change is occurring and is anticipated, output price must be falling. And this in turn implies a decline in the quasi-rent of capital equipment. This provides a connection between the bias of technical change, in terms of factor augmentation, and the rate of obsolescence of capital. Again Salter's 'vintage' model is used to analyse this relationship, from which it is concluded that an increased rate of technical change will tend to encourage capital augmentation.

Up to this point it has been assumed that the inventor is alone in the creation or possession of his invention. The issues that arise when there is competition between inventors are discussed in Chapter 5. In contrast to the normal case of competition in production, where it has the socially useful function of allocating resources efficiently, with technological competition there is a strong presumption that such competition would be undesirable. This is obvious where there is duplication of research activities. But even in the absence of duplication, which is the assumption in Chapter 5, there is a waste of resources in technological competition. These wastes had been pointed out by Sir Arnold Plant over half a century ago, and more recently by Yoram Barzel.

The framework of Plant's thinking is an important precursor of the modern theory of rent-seeking. And Barzel's analysis was the first to account for the dissipation of rent that occurs in the race to be first. The connection between rent-seeking and rent-dissipation is the fact that the set of problems thrown up by science and technology represents an open access pool of potential rent, which can be realised by patenting inventions that are truly the outcomes of combining research inputs with the state of science and technology. Open access competition draws socially excessive resources into research activity because science and technology are available at zero price. In the limit, the pool of rent is eroded away as research inputs are employed to the point where their average product equals their unit cost.

A necessary element of the rent-dissipation process is the ability of research resources to move from one field of rent extraction to another. Evidence that such movement takes place can be found in the empirical work reported in Chapter 7, where it is shown that the number of inventions produced in a field is inversely correlated with the level of activity (representing the demand price of inventions) elsewhere in the economy.

Both positive and normative analyses of the race to be first are presented in which the key constraint is the time–cost trade-off. Barzel's conclusions on the difference between competition and monopoly in the supply of new technology are confirmed. In addition it is shown that under either regime a shorter period of patent protection implies a lower present value of R & D costs and a longer timescale for invention projects. This points to the possibility of using patent duration as a means to control technological competition. The possibility of additionally employing the royalty rate as an instrument of public policy is also explored.

The final section of Chapter 5 examines the case in which there is a small number of competing research teams. This is modelled as a Cournot–Nash game with rent-seeking participants, research teams, whose research not only produces a probability of successful invention but simultaneously a chance of acquiring the patent. It is shown that there is an important difference between assuming that research production function is firm-specific and assuming that it is industry-specific. The latter case corresponds to the idea of open-access fishing for research results.

1.3 SYNOPSIS OF PART II

The main burden of the empirical chapters of Part II is to establish first, that invention actually does respond to economic factors, and then to interpret any such response. The econometric evidence on the bias of technical change reported in Chapter 4 is, on the whole, consistent with the notion that the observed bias is induced. That is to say, the overall tendency is for factor-saving biases to be correlated with relative factor price rises. The evidence presented and analysed in Part II reinforces the assumption that invention is, to a degree, determined by economic factors.

The single most important precursor on the subject of endogenous technical change is Jacob Schmookler. Chapters 6 and 7 rely heavily on Schmookler's work, and in particular re-examine the data that he painstakingly assembled. What is new in Chapter 6 are not the data, nor the attempt to establish whether economic time-series precede the patent time-series, where the number of patents is a proxy for the volume of inventive activity, but the method of analysis. The time-series approach to causality-testing originated with Clive Granger's (1969) contribution. This appeared three years after Schmookler's *magnum opus*, and by today's standards Schmookler's work appears distinctly informal. The graphical presentation of section 6.3 essentially reproduces Schmookler's argument that the synchronisation between capital goods patents and investment activity is more than coincidental. After summarising the philosophy of Granger causality, and how it is to be tested for in the present context, the empirical causality tests are presented in 6.5. The analysis does not significantly challenge Schmookler's own conclusions, though the results are mixed as section 6.6 notes. For some industries a significant causality relation is established, but for others the causality tests are non-commital. There is however no evidence for a causal relation in the opposite direction, from patents to investment. The conclusion of this time-series analysis is therefore that the data do support the hypothesis that invention is endogenous, although there is a further possibility that investment and invention are determined simultaneously.

Ideally, it would have been desirable to use a new dataset with which to carry out the causality tests. Unfortunately such data, spanning a sufficiently long period to apply time-series methods, are simply not available. It is of course second best to use Schmookler's data once more, but when data are scarce there is a good case for extracting their information in the most efficient way. Basically the same justification can be given for re-using Schmookler's cross-section data in Chapter 7. However, the novelty here is not just that more recently developed econometric methods are applied, but in addition the relationship between inventive activity (patent counts) and the level of economic activity (value-added) is given an economic interpretation.

It is argued in section 7.3 that Schmookler's regressions actually identify a supply curve of invention. The theoretical basis for this is the proposition of Chapter 3, that the demand price for an invention is proportional to the output of the invention-using industry, *ceteris paribus*. However, under this interpretation, it would not be correct to focus on the cross-section regressions, as Schmookler does. These appear to give an elasticity of unity, but since each industry has a very different technological base, this factor may be biasing the elasticity estimator upwards. It is argued that a more suitable way of estimating the overall elasticity of supply of inventions is to use the random coefficient regression method. Also, additional variables are introduced, representing, on the one hand, shifts in the supply function due to changing science and technology, and on the other hand, the influence of demand for inventive inputs elsewhere in the economy. The conclusion of this empirical work is that the elasticity of supply of counts of patented inventions is less than unity, and perhaps nearer one half. There is also substantial evidence that inventive resources can be attracted away from an industry by relative expansion elsewhere in the economy. Moreover, a highly significant partial correlation with total inventive activity elsewhere in the economy testifies to the fact that there are substantial common effects in inventive activity across industries, possibly reflecting the pervasive influence of scientific advance.

A different and complementary approach to the supply of inventions is taken in Chapter 8. Whereas in Chapter 7 the supply function is approached directly, as the *ceteris paribus* relation between changing price and volume of inventive activity, in Chapter 8 it is approached indirectly through the research production function. The key paradigm for research activity here is of research as a filtering or screening process. This way of looking at inventive activity, which reflects the analysis is Chapter 2, originated with Fritz Machlup. The set of potential research projects, and the set of potential research inputs or researchers, are not all of equal value. Research projects vary in terms of profitability, and researchers vary in terms of productivity. But if, before undertaking the research work, it is possible for the research administrators to form a judgement about the differing profitabilities of projects or productivities of researchers, the actual supply will be filtered. Projects will then be considered, and researchers employed, in sequence from the top end of the distribution downwards. It follows that the basic characteristics of supply, and in particular its elasticity, are determined by the perceived distribution of projects by net value and of inputs of productivity.

In order to assess the empirical implication of the filtering paradigm it is necessary to acquire information about these distributions. Of course, since they are perceived distributions they are not strictly observable. But it is assumed in Chapter 8 that perceptions correlate with reality. Now the task is to find out actual distributions. In order to do so, the literature was culled for all relevant evidence. It is not easy to come by. Consequently the data-gathering task was approached in an eclectic spirit, and the data presented in this chapter are assembled from a variety of sources. In view of the paucity of any kind of relevant information in this area, this approach seems perfectly legitimate. But it does mean that the reader should see this as a synthesis or overview. The conclusions are drawn not on a statistical examination of a single dataset, but on the weight of evidence from various and disparate sources. It should be recognised that variety, or cross-validation in statistical terms, is a strength if one independent set of data reinforces the impression gained by another.

The final section of Chapter 8 attempts an assessment of the severity of diminishing returns in the invention-production function by identifying the relationship directly. It reports the results of a regression analysis of counts of internationally patentable inventions against research inputs in a cross-country and inter-industry setting.

Pulling all the strands together regarding the responsiveness of invention to the economic variables that affect its profitability, analysed in Chapters 7 and 8, it is clear on the one hand, that there is a supply response, as Schmookler had originally observed. But that response also seems to be rather inelastic.

The volume of inventive output is measured in Chapters 6–8 in terms of patent numbers. Since inventions differ in importance, a better measure is the rate of cost-reduction, or productivity increase, that they give rise to. Of course there is a lag, and not all measured cost-reduction is ascribable directly to invention. The Schumpeterian sequels of innovation and imitation are also important. Nevertheless, the social value of invention can only be assessed in terms of its economic effects. However, in this book there is no attempt to use the rate of productivity-increase or the rate of unit cost-reduction as a measure of inventive output, mainly because this was the approach applied and reported earlier in Wyatt (1983). But it will be useful briefly to summarise how that work relates to the present book.

The monograph, *Multifactor Productivity Change in Finnish and Swedish Industries, 1960 to 1980*, reports an attempt first to measure the rate of technological progress, and then to assess the marginal product of research capital on the basis of a model that imputes part of the progress to research and development. The rate of technical progress was measured as the rate of change of total

factor productivity. If a production function has constant returns
to scale then total factor-productivity change is simply the negative
of the rate of unit cost-reduction. The theory of productivity
measurement has been refined in recent years due to advances in
production function analysis and its connection with the theory of
index numbers. The monograph made use of these theoretical
advances.

Previous measurement of total factor productivity had tended to
be at a highly aggregative level. But from the point of view of
invention, it is more interesting to measure it at a more disaggregated industry level. Such disaggregated productivity estimates
have been calculated for the United States, but not elsewhere so
far. In the application reported in the monograph, total factor productivity changes were calculated for two-digit industries in two
countries: Finland and Sweden. So far as is known, this was the
first international inter-industry comparison attempted. The
reasons for this are first, that the data in many countries are not
sufficiently well developed for an industry analysis (an example
here is the UK, where gross output price indices are deficient and
capital stock estimates are inadequate for a number of industries);
and secondly, that it is exceedingly difficult to find pairs of countries with well-developed data in which the data are reasonably
comparable.

The monograph suggests that the social rate of return to private
research capital is very high. It seems, however, to have fallen
between the 1960s and 1970s in both countries, and there is a hint
that it has been higher in Finland than in Sweden over both
decades.

The model within which these tentative results were established
is standard, and based on the neoclassical paradigm. It imputes
research capital on the basis of zero depreciation, and assumes that
the marginal product of research capital is equalised across
industries.

The latter assumption obviously implies that research resources
are substitutable between industries, which proposition is certainly
consistent with Schmookler's views, and the results reported in
Chapter 7 here. It is not difficult to establish that a log-linear model
based on these assumptions implies that the rate of unit cost-
reduction varies linearly with research intensity, and that the coefficient of this relationship is actually the marginal product of
research capital.

There are, of course, many possible reasons why the rate of return to research capital might be high. Among them are the uncertain nature of research activity; the fact that social and private rates of return might diverge substantially due to the difficulties of privately appropriating the benefits of research; and the possibility alluded to earlier, that the supply of research resources is price-inelastic. In this sense the monograph provides indirect support for the inelasticity conjecture arising out of Chapters 7 and 8 of the present work.

1.4 SOME TENTATIVE POLICY IMPLICATIONS

The economic analysis of invention does not, at this point in time, yield unequivocal conclusions. But it would be wrong to give the impression that nothing of relevance to policy can be said. The key issue is the degree to which invention is endogenous. Here the main conclusion is ambiguous. It is that inventive activity does indeed respond to economic rewards but, due to very sharply diminishing returns to research, it seems that the value of inventive output may be much less responsive.

The fact that inventive activity responds elastically to the demand price of invention implies that it can be channelled in the right direction by suitably designed incentives. They should be designed to make the private rewards proportional to the potential social value of inventive output. It is precisely this that the patent system achieves, though of itself this does not vindicate the creation of monopoly that it implies. While the direction of research thus encouraged may be roughly correct, it probably also has the effect of drawing socially excessive resources into inventive activity as a whole, and directing too much within the total towards patentable inventions. There is also the deadweight monopoly loss once the invention has been produced to set against the inducement advantages of the patent system. The analysis in this book suggests that induced inventions will typically be of minor importance, representing improvements to existing technologies perhaps, rather than fundamental changes.

It can be predicted that changes in the rewards for invention will imply an elastic response in terms of inventive activity or the number of inventions produced. Thus if new categories of technology become patentable, researchers will redirect their energies accordingly. Recent corroboration of this has been seen in the response to the change in United States patent law that allows new hybrid species of plants to be patented. But it may reasonably be doubted whether such changes are important for the production of socially valuable technological change. Most induced inventions will be of minor importance.

Another avenue of public policy towards technical change is the idea of subsidising research activity, or giving special tax allowances. If such assistance is directed to specific sectors or types of technology then it can be predicted that these sectors will mainly draw research resources from other research activities. There will indeed be more inventive activity in the favoured sectors, and more inventions produced there. But on the whole they will represent minor improvements only, and it is doubtful that the overall social value of technical advances will exceed that which would have obtained in the absence of the special inducement. Indeed, it is more likely to have diminished if the original allocation of research resources was in line with the demand price of invention.

If all research activities are favoured by special tax or subsidy treatment, then resources will be drawn from the rest of the economy, and there will be an overall increase in technical change. But the sharply diminishing returns to inventive activity suggest that the response is likely to be small. This is not to say, however, that such a policy would be inappropriate – that would depend on the usual welfare judgement comparing marginal social and private values.

A different type of public policy towards research and invention is public spending on science. The social value of new scientific results is not normally privately appropriable. The output from spending on science will tend to create new opportunities for applied research and will tend to lower the unit cost of inventive activity generally. Both effects will stimulate applied research and inventive activity, and if scientific advance operates to reduce the marginal cost of invention, that would loosen the critical elasticity and could have a significant impact on technical change. For example, scientific advances may make it easier for the marginal researcher to go about his job, by providing easier techniques or more systematic methods. Or perhaps the scientific principles become clarified. Very productive researchers may have mastered these aspects anyway, but the less productive researchers now find it easier to produce the results they are after; and, according to the analysis of Chapter 8, a significant part of the inelasticity of supply of technological improvement is due to the supply of effective inventive labour.

It was mentioned at the beginning of this chapter that the main focus of this book is on the economic determinants of inventive activity. Neither the consequences nor the desirability of invention are examined in the following chapters in any detail. It would therefore not be appropriate to dwell here on possible policy implications of the analysis. The tentative thoughts that have been briefly outlined in this section are obviously only meant to be suggestive. It is to be hoped that others will pick them up and give them the attention they deserve, for these are important and difficult matters. It should be clear from the foregoing that a matter of primary importance in future research will be to establish with some confidence the proposition on which the comments in this section are based: namely, that the supply of technological improvements, as opposed to numbers of inventions, responds rather inelastically to demand inducements.

2 Research and Inventive Activity

The aim of this chapter is to explore the nature of inventive activity in order to establish, if possible, some predictions about its economic characteristics. The classifications of research activities used in official surveys or by economists who have studied this subject are examined for possible insights into research as an economic process. One such classification distinguishes between agenda-increasing and agenda-reducing research activities, and this is considered in section 2.2. Another classification is between sequential and parallel programmes of research, which are studied in section 2.5. This chapter also makes use of the fact that research output is essentially information, and explores various paradigms that this analogy gives rise to. For example, the idea that research is a kind of search activity is taken up in section 2.3. The aim in all of this is to be able to say something about 'the research production function' from an *a priori* standpoint. In other branches of economics it would be just as useful to examine the 'supply function', and this alternative approach is explored in section 2.4, which finds an empirical counterpart in Chapter 7 below.

2.1 RESEARCH AND INVENTION

Inventive activity absorbs a substantial amount of economic resources. Officially defined research and development expenditures amount to about 3 per cent of GNP in most advanced countries, but not all inventive activity is caught by official definitions and measurement. Although it is possible to affirm the scale of this activity, it is much more difficult to define what inventive activity is. This difficulty is due to the essentially problematical nature of the output of inventive activity. It may be an invention – an addition to the set of blueprints for products of productive processes – or more generally and less concretely it may be an advance in knowledge or an increase in information. The output is essentially problematical for a number of reasons. First, it relates to problem-solving – even if, as is often the case, the problem was not posed explicitly in advance. Secondly, the output of inventive activity cannot be fully anticipated, for if it were it would represent a mere reproduction of old knowledge; and thirdly, for the same reason, it must be unique so that all outputs from inventive activity are distinct or heterogeneous. Fourthly, it may well be that there is no identifiable output: research is characterised by such blind alleys, but even apparently fruitless research projects contribute to knowledge so long as the research activity itself is original.

It is clear therefore that the output from inventive activity or research – the terms will be used interchangeably – is not easily defined, let alone measured. It is also apparent that these problems of definition and measurement inhere in the topic and are additional to the problems that occur because research is a 'service', for which outputs are measured by inputs. For most services, even bureaucratic services, it is possible in principle to define the output of that activity and hence in principle to measure it. The irony will not be missed, therefore, that it may not be possible to measure the productivity, or the change in productivity, of an activity that is widely believed to be the principal cause of changing productivity elsewhere in the economy.

Definitions of research and invention tend to be formulated with the objective of unambiguous classification and measurement in mind. Thus it is the collectors and assemblers of statistical material in this area that provide the most widely quoted definitions: organisations like the National Science Foundation in the United States and the Organisation for Economic Cooperation and Development (see NSF, 1972; and OECD, 1975, for example). However, commentators using these or similar definitions point to their often arbitrary derivation or implications, and also to their often fuzzy distinctions, e.g. as between 'applied research' on one hand, and 'development' on the other.

It is, of course, easier to define an invention than it is to define inventive activity. The common definition of an invention is grounded in patent law as being 'a prescription for a new product or process that was not obvious to one skilled in the relevant art at the time the idea was generated'.[1] However, Siegel's (1962) attempt to distinguish invention from discovery leads him to a rather tortuous definition of inventive activity, as follows: '[it] may be regarded as purposeful and practical contriving based on existing knowledge (theoretical or applied) and uncommon insight or skill; that is as the art of bringing to workable condition a potentially economic or usable process or product ... that has a significantly novel feature.'

The problems of suitably defining the output of research do not, however, vitiate exploration of different ways of looking at the research process nor examination of the implications of these analogies or paradigms. Like the elephant that is more easily recognised than defined, research and inventive activity can be discussed on an implicit understanding of the subject-matter. This approach seems more productive than a dissection of alternative definitions.

It has been implicit in the foregoing that the output of research activity is not always – or even usually – an invention. Consider, for example, the common categorisation of the research process as an ordered spectrum from 'basic' or 'pure' or 'scientific' research at one end through 'applied' to 'development' at the other end. It is apparent that at each end of the spectrum, the term 'invention' is unlikely to be appropriate. When a pure scientist doing basic research notes some regularity in nature he may be said to *discover*, not invent, a law. And a development engineer will *contrive* or *design* an improvement to a prototype. As Fritz Machlup (1962) puts it:

To describe the difference between basic and applied research, it has been suggested that the former is after discoveries, the latter after inventions. There is much to be said for this suggestion; as I see it, the concepts of 'inventive activity' and 'applied research' overlap to a large extent. (p. 148)

In a later section of the same chapter Machlup dissects the semantic and logical distinction to be drawn between discovery and invention. He concludes that: 'we discover, by means of our faculty of perception, what has been there before, though unnoticed; we invent, by means of our faculty of mental construction, something new and of importance, scientific, artistic or technical' (pp. 163–4). In this usage it is, of course, possible to invent a scientific theory. Indeed, Machlup avers that the 'widely held opinion that "laws of nature" are discovered, not invented, is rejected nowadays by most philosophers of science.'

The distinction between 'invention' on the one hand, with its connotations of purposeful contriving, and 'discovery' on the other, with its overtones of serendipity or finding by accident, is suggestive but in the end not entirely satisfactory. There may be, and perhaps even usually is, an element of discovery in invention. This is captured by the phrase often used in a patent law definition of invention, that 'it should not be obvious to those well-versed in the appropriate industrial arts'. But, beyond this, there is a convincing argument that some passive process of discovery is an important ingredient in many forms of creativity. The case is presented by Arthur Koestler (1964), who associated creativity with unconscious mentation, or dreaming, in which the mind discards logic and instead makes lateral connections between the problem as formulated in its conscious mode and other elements of reality or imagination. Thus inventions may often be the outcome of a combination of active search and passive discovery.

Koestler discusses a number of paradigms of creative activity. His preferred paradigm is geometric: new ideas arise from the intersection of what were thought to be logically separate matrices. It is interesting to note that William Shockley, himself an inventor of considerable distinction, came to a similar view when he considered the causes of the extreme variations in researcher productivity which are discussed in Chapter 8 below. Shockley sees invention arising from the concatenation of distinct modes of thought, and argues that the ability simultaneously to manipulate several distinct frames of reference is a rare gift in people, which is what accounts for the widely different performances of research scientists.

Reference has already been made to the categorisation of research into 'basic', 'applied' and 'development', which is often employed in official statistics. When this taxonomy is seen as an ordered spectrum of distinct processes through which any given invention might pass, it is referred to as the 'linear model' of the research process.

The linear model attempts to impose a simple-minded causal sequence on the categorisation. It is clear, however, that applied research and even development can throw up problems which lead to basic research and, equally obviously, inventions can arise at any of these 'stages'. Perhaps the attraction of the linear model arises from the fact that practically all inventions build on previous results and experience in science and technology, so for each it is possible to trace back its ancestry and precursors; and inevitably a new technology will at some point use basic scientific principles. This type of geneology cannot however justify the linear model because, as with all family trees, those for inventions tend to have branches too numerous and interwoven to be unravelled into the single strand that the linear model assumes.

This notwithstanding, the simple tripartite classification of research into 'basic', 'applied' and 'development' is usefully correlated with a number of economically important features of inventive activity. Invention, both as an activity and as the outcome of the activity, has a number of remarkable economic characteristics. Most importantly, it combines several properties of public goods with uncertainty. Thus, unlike 'normal' goods, inventions typically have strong externality features, are difficult to appropriate, and are characterised by extreme indivisibility or 'jointness in supply'. All of these public good characteristics become increasingly severe at the 'basic' end of the research spectrum, as does the fundamental uncertainty inherent in the research process. Thus basic research can lead to advances in science, which may affect the cost of a wide gamut of other research and production activities (externalities and jointness); but such advances are typically difficult to appropriate – patents, for example, cannot be taken out on 'scientific laws'.

At the other end of the spectrum, the research activity described as 'development' is much more like a private good. Even if the results of development research are not exactly predictable, at least there is a vastly reduced range of possible outcomes. Uncertainty is thereby more manageable, and along with that the specificity of the research outcome is much greater. There might be limited externality effects, and the product is in all likelihood appropriable, and even possibly marketable because perhaps it is literally a (new) product, or if it is an idea it can be patented.

It is not surprising, therefore, that the bulk of basic research is carried out in the public sector, nor that the preponderance of resources devoted to inventive activity in the private sector is aimed at development. 'Applied' research may supposedly occupy the middle ground, though it is argued in section 2.3 and in Chapter 8 that one of its functions may be as a filter or screening process whereby the potential research projects arising at the basic end of the spectrum are weeded out to improve the average profitability of development projects.

But while there may be a correlation between the tripartite basic–applied–development classification and some fundamental economic characteristics of research, the classification is ultimately arbitrary and not of great assistance in the economic analysis of invention. To illustrate its arbitrariness, consider the activity of designing to meet functional rather than aesthetic objectives, which is a major use of the time and effort of a large number of engineers and others. In its essential characteristics of novelty, originality, contrivance and so on, design resembles inventive activity. Indeed, many inventions arise from a 'design need'. More than this, the products of design work, blueprints, have essentially the same economic features as inventions. They lead to the original mould of a physical product or productive process. Economic analysis of designs and inventions must recognise their uniqueness. In this they are very different from articles of mass production, which is more properly referred to as 'mass reproduction', an expression coined by Nicholas Rescher (1980). But the view of most people is perhaps that designing is a more mundane activity than inventing, although in essence they are similar. Much research work might just as well be described as designing (indeed we speak happily of the 'design of experiments'), yet there is little doubt that the bulk of design work is excluded from the official statistics which use the basic–applied–development trichotomy.

2.2 INFORMATION AS THE OUTPUT OF INVENTIVE ACTIVITY

To insist that the output of research is an invention or a discovery is to ignore perhaps the vast majority of inventive activity. For it predicates, unjustifiably, a successful outcome of the activity. It may therefore be better to consider the output of a research process to be knowledge, or rather an addition to knowledge. The sense in which 'knowledge' is used here corresponds more to impersonal 'information' than to 'wisdom' which is invested in an individual. Thus while a research project may not have any clear outcome such as an invention, its negative conclusion will nevertheless have contributed to knowledge.

An incomplete characterisation of the research process as knowledge creation is to see it as a search for the true 'state of nature' from a given set of possible states of nature. Thus if ignorance is the converse of knowledge, it may be supposed that ignorance implies uncertainty about the true state of nature. And the generation of knowledge by research has the effect of reducing this uncertainty.

The transposition from knowledge/ignorance to certainty/ uncertainty about the true state of nature implies the possibility of degrees of knowledge and ignorance, since one can be more and less uncertain about the truth of a particular proposition. In this way the black-and-white nature of the contrast between ignorance and knowledge can accommodate a spectrum of intermediate greys in an uncertainty scale. More than this, the translation of knowledge into certainty and ignorance into maximal uncertainty also suggests the possibility of a measure on that scale: it would naturally be related to the concept of probability as a degree of belief.[2]

Before continuing on this theme, note that the paradigm was stated to be an incomplete characterisation of the research process. It corresponds to what Machlup (1960) describes as 'agenda-reducing' research activity. The process consists of whittling down or eliminating elements from the given set of possible states of nature. But there is another aspect to research as knowledge creation, which is to do with determining the elements in the set of possible states of nature in the first instance. This is what Machlup calls 'agenda increasing' research activity. Thus agenda-reducing research consists in solving problems, whereas agenda-increasing research consists in finding interesting problems to solve. They are complementary rather then contradictory elements of the research process.

Even if the solution to a problem is not known it may be possible none the less to say that it must lie within one of a finite number of mutually exclusive possibilities. In fact, unless such an exhaustive and mutually exclusive categorisation is implicit in the statement of the problem it is possible to argue that the problem itself is not well posed. Now even the most complex set of possible solutions can be reduced to a series of primitive solution sets in each of which the answer to the subquestion must be either 'yes' or 'no'. Furthermore, if the context of the problem ensures that one of these answers must logically follow from the given data, then the problem is insubstantial since it is deducible. For such problems the answer is implicit or encoded within the statement of the problem, and simply requires computation. By contrast however there are problems for which no amount of computation power of logical deduction will provide *the* answer. For these there is always some residual doubt or uncertainty which is inherent in the problem, and the doubt will only be resolved by new evidence or a novel approach. That is, the uncertainty can only be eliminated, or even merely reduced, by changing the given data. Such problems require methods of *inductive* rather than deductive logic, and it is to problems such as this that the term 'inventive activity' best relates, carrying with it its various connotations of contrivance and novelty.[3]

Suppose a draughtsman wishes to choose a type of pencil for his work, not having used one before, and suppose that the only element of choice is in the cross-sectional shape of the pencil, of which there are just two possibilities: circular and hexagonal. Assume that in all other respects the two types of pencil are similar, and in particular that the prices of the two are equal and not envisaged to change. The draughtsman's problem is one of utility-maximisation under uncertainty – the uncertainty being in the utility function itself. With no further data to guide his choice, his uncertainty, or lack of knowledge, is as large as it can be, and his decision between the two pencil types will be arbitrary. So the odds of choosing either are equal. But the uncertainty may be reduced, or knowledge about the problem increased, by bringing in more data. One way might be by careful consideration of the essential functions of the implement – for example, the necessity of precise lines may imply a need for a sure grip and the further implication that the hexagonal shape might be preferable. Another way of introducing more data might be through analogies with other similar implements like pens. Of course, yet another way would be to test both types. These new data may partially resolve the uncertainties: for example, the odds in favour of one type being superior for the job may have increased to 10:1 after a brief trial. By this time the draghtsman might have effectively discovered (invented for himself) the right tool.

Notice how the draghtsman's problem is essentially inductive, and that this follows from his initial ignorance. By contrast, the standard formulation of the individual consumer choice problem, that of maximising a deterministic utility function subject to a budget constraint and given substitution possibilities (price ratios), is essentially deductive. The draughtsman's inductive problem requires him to step outside the framework of the problem and either reformulate it or else bring in new pertinent evidence. The solution to the standard consumer choice problem, by contrast, follows logically from the premises.

For the draughtsman, the possible states of nature are: (i) circular is superior for his purposes, and (ii) hexagonal is superior. His ignorance about the true state of nature assigns an equal probability to each of the possible states as being the true one. His personal research work (i.e. bringing in new data and reformulating the question) changed the probability assignment until eventually a sufficiently large proportion of the probability mass was assigned to one of the possibilities that a decision could be made as to the appropriate design for his purposes. This indicates that the more evenly spread out is the probability assignment over the possible states of nature the more uncertainty there is as to the true state of nature. We may say that the larger the *entropy* of the probability distribution, where entropy H is defined as

$$H = - \sum_{i \varepsilon \{i\}} p_i \ln p_i$$

the greater the uncertainty. In this formula i indexes the possible states of nature, the complete set of which is $\{i\}$, and p_i is the degree of belief probability that the ith state is true.

Now, all probabilities are conditional, and the draughtsman's research or inventive activity regarding the appropriate design of the pencil involved changing the conditions under which the probabilities were assigned to the various possible states of nature. The objective of this research process is now understood to be a reduction in entropy, for that is implicit in the resolution of the probability mass towards a particular state of nature.[4]

This form of invention is the search for new data to decrease the entropy of the subjective probability distribution over possible states of nature. Let D represent the given data for the problem, then entropy may be written:

$$H(\{i\}|D) = - \sum_{i \varepsilon \{i\}} p(i \mid D).\ln p(i \mid D)$$

It is supposed that the research task consists in changing the conditional data D to D′ such that the decrease in entropy:

$$\Delta H = H(\{i\} \mid D) - H(\{i\} \mid D')$$

is maximised. In Machlup's (1962) terminology, this is 'agenda-reducing' research activity.

Define the 'equivalent number of categories [states of nature] of the distribution' as:

$$N = e^H$$

When the distribution is quite flat then H, entropy is at its maximum value, ln(n), so $N = n$, the actual number of categories or states of nature. By contrast, when the distribution is strongly peaked, then $H \approx 0$ and $N \approx 1$. With this definition it seems reasonable to interpret agenda-reducing inventive activity in the following precise manner: it is the kind of inventive activity which, for a given set of n states of nature, aims at sharpening the probability distribution over those states in the sense of reducing the equivalent number of states of nature, N.

But while it is possible to be precise about what is meant by agenda-reducing research activity, and even in principle how to measure it, no such possibility is obvious for agenda-increasing research. Thus, as Griliches (1973) points out, if invention is like fishing red balls from an urn full of white and red balls, it is straightforward to conceptualise. But if it also consists in finding new urns from which to fish, which had previously been hidden, that is a different matter. Some kinds of research are essentially agenda-reducing, however, and for these the information theory-based measures such as entropy reduction might well provide useful guidance in practical decision-oriented analyses of research output. Thus experiments designed to narrow down the uncertainties of an already sufficiently well-defined problem, as for the archetypal draughtsman, might be analysed in that way. But, being limited to only part of the research process, it would seem that a suitable measure of research output for general descriptive or analytical purposes must be sought elsewhere.

It is reasonable, therefore, to be sceptical about the possibilities of information theory to provide a measure of research output, despite the fact that such output is agreed to consist in new knowledge. Information theory can provide a measure of the amount of information, like reduction in entropy, contained in a 'message', but the value of the message is also determined by its actual content – not only by how surprising or predictable it was. An information theory approach might be perfectly adequate if all information could be encoded as yes/no answers to a sequence of precise questions. This is pure agenda-reduction. But information theory has no place for messages that are wholly unanticipated in the sense of bearing no relation to prior questions. These increase the agenda of possibilities. New knowledge is not synonymous with increased information.

2.3   PARADIGMS FOR THE PROCESS OF INVENTION

This section acknowledges the essentially negative conclusion of section 2.2, namely that a natural measure of the output of individual research projects cannot be derived from an analysis based on first principles of the nature of invention. Nevertheless, there are some useful paradigms of the invention process in the large that do have a bearing on the supply of inventions. These are mainly based on a view of invention and research that assigns the primary role to its chance or haphazard nature. Thus research is seen as a process of search in an uncertain environment. The question then is: how might rational actors be expected to behave?

Several paradigms of the research process present themselves, largely by analogy with the way that decision-makers are thought to respond in the face of risk. In other branches of economics it is often assumed that people are risk-averse, or at most risk-neutral. Thus a parallel can be seen with processes of hedging and risk-spreading, such as insurance schemes or portfolio selection. On the other side of the coin, there is usually expected to be a reward to individuals who take on risks. This is, of course, related to the activity known as entrepreneurship. Elements of all of these can be visualised in the process of invention.

To see elements of entrepreneurship in invention would of course surprise someone schooled in the Schumpeterian tradition. In the first place, for Schumpeter's classification of the sequence invention–innovation–imitation, the initial phase of invention is exogenous. It is a datum to which the economy reacts. Entrepreneurship is associated rather with the innovation phase. And here, secondly, it is the aspect of charismatic leadership rather than risk-bearing that Schumpeter emphasises.[5] But Schumpeter's conception of exogenous invention cannot be accepted. Invention undoubtedly responds to varying rewards as, for example, the work of Jacob Schmookler[6] testifies, and which is confirmed here in Chapters 6–8. As to the definition of entrepreneurship having to do with qualities of charismatic leadership, which entails 'the character, courage and, above all, vision, required to depart sharply from accepted routines and practices' (Rosenberg, 1976, p. 67), this might indeed be another way of describing risk-taking. It is perhaps normal to consider a risk-taker as someone with an unusual utility function over uncertain outcomes. But it would be equally valid to see him as being unusually gifted with vision and the character to pursue it. In other words, he has a clearer idea of the risks and opportunities, even if his utility function is not in any way unusual. This would be nearer to the Schumpeterian conception of entrepreneurship. It is not, however, confined to the act of innovation, but applies also to inventive activity too.

One difficulty with the idea of entrepreneurship in inventive activity is the apparent fact that as technical change has gathered pace over the past century, the role of the individual inventor seems to have declined. Entrepreneurship is, after all, more a quality of individuals than of organisations. On this there are a couple of points to be made. First, as the important study of John Jewkes *et al.* (1960) shows, individual inventors are still disproportionately represented among important inventions. Secondly, inventions are made by individuals or small groups of individuals even if they are employed in a larger research outfit. These individuals may indeed assign their inventions to their employer, but it does not follow that they go unrewarded for their special output. They have traded their uncertain entrepreneurial income for a salary and in return the employer has assumed a larger role in bearing the risks and gathering the rewards of their activity. That is to say, the commercial research laboratory may be a more efficient vehicle for inventive activity by virtue of its greater ability to contemplate risky research, due in part to a spreading of risks across a portfolio of projects.[7] The laboratory's 'utility function', in other words, makes it less risk-averse for particular projects, but the Schumpeterian qualities of charismatic leadership in the sense of vision and the character to pursue it are still vested in individual researchers.

There is a difficulty with the notion that in research there can be economies of scale in the sense of risk-spreading. The idea, which underlies both insurance and portfolio selection, is that by pursuing a number of projects the overall risk, measured by the variability of the outcome for a given expected return, can be decreased. This has, of course, stronger validity the less the individual projects are correlated. But even if the projects are independent and stochastically orthogonal (zero correlations between their outcomes), the proposition that the variance of a combined outcome is less than that of the separate projects for any given expected value only holds for a limited set of types of population probability distributions. And, in particular, the second moment of the distribution must be finite. But this is not true of all distributions; for example the variance of a Cauchy distribution is infinite, and for the class of stable Pareto distributions it is only finite if the exponent parameter $\alpha \geq 2$. Unfortunately, there is evidence[8] to suggest that the returns to inventions appear to be distributed *à la* Pareto with $\alpha < 2$. This implies that the variance of outcomes may actually be increased if more projects are brought under the wing of the research institute.

It is useful in this regard to separate out two distinct classes of invention, namely 'run-of-the-mill' inventions and 'spectacular' inventions.[9] Numerically, the vast bulk of inventions are of the first, run-of-the-mill variety. But though spectacular inventions are the exception, they are the focus of the Schumpeterian vision. The arguments and empirical data presented in Chapter 8 below support the idea that run-of-the-mill inventions are endogenous or induced by economic circumstances, whereas spectacular inventions are exogenous. It is this exogenous component in inventions that accounts for the characteristic tail of the Pareto-like distribution of inventions by value. They are more like discoveries than the result of inventive search and contrivance. When they are extracted from the set of inventions and put to one side, what is left is much more of a standard frequency distribution by value, and one to which analysis of risk, search and portfolio selection may be applicable. It is assumed in what follows that search processes apply to these run-of-the-mill inventions.

The very word 'research' seems to call for some paradigm of search by which to describe it. Indeed, Chamber's Dictionary (1898) defines research as: 'A careful search: diligent application or investigation: scrutiny'. And different search paradigms can be applied to research and inventive activity. Among these are the paradigms of sampling, filtering and fishing. The term 'fishing' is used here to denote a process of competitive search, which gives rise to particular economic problems when there are open access 'fisheries'; this aspect of search is deferred to Chapter 4 where the theme is competition among inventors.

By 'filtering' is meant a process of selection and choice from an agenda of research projects, say. Suppose, for example, that a research unit has a large number of possible projects that it could pursue. It might have a prior but diffuse notion of the expected returns from these projects. With a limited budget it must choose a subset of projects upon which research can be carried out – where the process is 'agenda-reducing research', as defined in section 2.2. If all projects look the same, *ex-ante*, in terms of expected returns and risk, then choice of a subset of projects would be random and the research unit can expect the average return only. But if the researchers are able to rank projects, they will work down the agenda from the most profitable and can expect a greater average return by so doing. The preliminary research that sharpens up the prior distribution of projects by expected value and produces a ranking of projects is a process of filtering. It seems plausible that much 'applied' research, before projects get to the 'development' or testing stage, is a form of filtering. This process is dicused further in Chapter 8.

A basic distinction here is between prospects with economically independent outcomes and those for which the outcome of one research project is a substitute for that of another project. The standard theory of portfolio selection, along with the filtering process referred to above, serves as an adequate model for projects with economically independent outcomes. That is to say, a rational selection of a portfolio of such projects can be based on their expected returns and co-variances, where the latter, if non-zero, are determined by the demands for the outcomes, not the technological interrelations between the projects at the research stage. A project is thus carried out if it belongs to the 'efficient set', in which it is not dominated in terms of risk and return by any combination of other projects. The overall expected value of a set of such economically independent projects is simply the sum of their individual expected values. And if the decision-maker is risk-neutral with an unlimited budget, all projects whose expected value exceeds their costs will be carried out. With a limited budget, projects will be carried out in order of their expected benefit/cost ratios until the budget is exhausted. With risk-aversion, the set of projects selected will be the one on the efficiency frontier for which expected utility is highest.

The sampling, as opposed to filtering and selection, paradigm applies where one project's outcome is a substitute for that of another project, or where there is a choice between projects with the same end result. There may appear, *ex-ante*, a number of different ways (research projects) of achieving any one of the economically independent prospects referred to in the previous paragraph. Because the goal is the same, these projects are technological substitutes for each other. Two very different strategies are open to researchers faced with a set of such substitute projects. In one the researchers select a set of projects and pursue them simultaneously. This is known as *parallel research*. In the other strategy the projects are taken on one at a time. This is *sequential sampling*. Where there is a single end result 'success' or 'failure', the sequential strategy stops with the first success. With parallel research there is a possibility of redundancy in the form of several successful outcomes. It may often be the case that the successful outcomes, though substitutes, are different and have varying economic values. In this case both the parallel and the sequential sampling strategies are looking for the maximum expected return.

As an illustration of a sequential sampling process, consider the techniques of statistical process control known as EVOP (evolutionary operations).[10] This applies to the search for the optimal way of carrying out an existing industrial process. It is apparently very successful for chemical processes where several parameters such as temperature, pressure, viscosity and so on are controllable. Strictly speaking, EVOP is a systematic procedure for optimising the running of existing facilities, and as such is an aspect of learning by doing rather than invention, though some modified analogy may apply to the operations of a research team. Figure 2.1 illustrates the principles by which EVOP works. The contours show the different yields that can be achieved by appropriate combinations of the control variables, temperature and pressure. The yield contours are unknown to the investigators (researchers). Current production is represented by point P, on which improvements are sought. The EVOP procedure consists in selecting some neighbouring points to P, marked by crosses, and running the process at those points on succeeding days (or hours, etc.). The points are selected randomly, and sufficient repetition is allowed until it is clear which direction, if any, represents an improvement. Obviously the point north-east of P is the best, and this provides the base P′ for the next round of search. Eventually the process con-

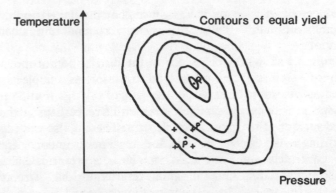

Figure 2.1: Evolutionary operations

Source: Adapted from Box and Draper (1969), p. 164.

verges on the optimum, at Q. But if the sampling is expensive the procedure should stop short of the optimum shown in the figure, where such costs are ignored.[11]

Most inventive research cannot be described like EVOP as a process of systematic sampling by factorial designs in a strategy space. Instead, a more primitive search paradigm is needed. The literature on search in economics originated with George Stigler's (1961) essay on search for the lowest price in a market with price-dispersion. If it is maintained that the subjective probability distribution of values from which researchers select projects is stable, then techniques similar to those employed by Stigler can be brought to bear on the matter. In this conception, the entities of interest are the largest order statistics of samples of various sizes from the relevant distribution.

If a sample of n drawings is taken from a probability distribution with cumulative F and density f, then the probability density g of the maximum (the *n*th order statistic) is given by:[12]

$$g(u) = n.f(u).F(u)^{n-1} \qquad\qquad 2.1$$

where the random variable u is the largest realisation in a sample. If it were known what distribution characterises research sampling, the corresponding function g(u) could be used to develop an economic theory of research and invention according to the search paradigm. Unfortunately, of course, this is not known, but in the remainder of this section it is asumed that the underlying distribution is normal, initially with mean zero and unit standard deviation.

Figure 2.2 shows the behaviour of the distribution of the maximum of samples from a N(0,1) distribution as sample size is increased. As would be expected, the mass of this distribution shifts rightwards with larger samples. Less intuitive perhaps is the reduced spread with increased sample sizes. There is also evidence of increasing left-skewness; but this does not appear important, even with moderately large samples. What is clear, however, is that there are strongly diminishing marginal returns to this stereotyped sampling research.[13] Both the increasing expected value and the decreasing standard deviation change much less than in proportion to sample size as sample size increases.

In fact, if attention is restricted to the behaviour of the expected value of the maximum, which would be appropriate for a risk-neutral decision-maker, it follows a simple yet unexpected function of N, sample size, namely:

$$E(u(N)) = a + b.\sqrt{\ln N} \qquad\qquad 2.2$$

This relationship is empirical rather than analytical,[14] but as Figure 2.3 shows, it is very accurate for suitably chosen constants a and b. The expected marginal product of sampling, or the increment to expected value of the maximum as sample size is increased by one, can be written:[15]

$$E(\text{marginal product}) = b/(2N.\sqrt{\ln N}) \qquad\qquad 2.3$$

Assuming that the research unit chooses sample size so as to maximise expected value net of sampling costs when parallel projects are undertaken, and that incremental sampling costs are constant, the relationship 2.3 can be inverted to give optimal sample size N^* as a function of unit sampling cost, where unit sampling cost equals the expected marginal product of sampling by virtue of the maximisation assumption. This is, in other words, the derived demand for sampling as a function of the incremental cost of sampling. Within the research paradigm it can be interpreted as the derived demand for research as a function of the cost of research. In this context it is interesting to note that this derived demand for research is inelastic for all sample sizes. The elasticity η is:

$$\eta = -2.\ln N^* / (1 + 2.\ln N^*)$$

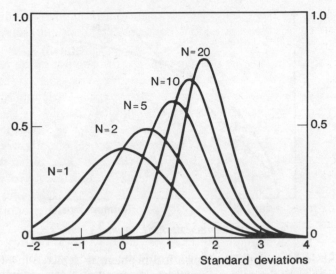

Figure 2.2: Distribution of maximum from normal samples

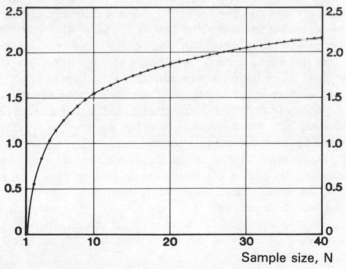

Figure 2.3: Expected value of maximum from standard normal samples

Note: The equation of the curce is:

$$Y = -0.783 + 1.5319 \sqrt{\ln(N)}$$

and was estimated for the 36 points, N = 5 to 40

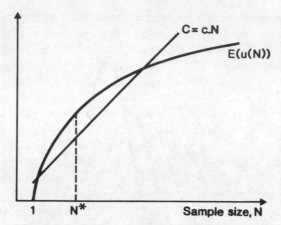

Figure 2.4: Determination of optimal sample size

The research unit's optimisation problem is depicted in Figure 2.4 which displays the expected benefit function E(u) and the cost function $C = c.N$. Optimal sample size is given by the condition that these functions have the same slope, at N^*. Obviously, as the slope of the cost function, which represents the marginal cost of sampling, c, rises so the optimal sample size falls. Also, increasing fixed costs will shift the cost function vertically upwards. It can be seen that this would leave N^* unaffected so long as any sampling research at all is undertaken – but of course it may affect the decision whether or not to engage in sampling research in the first place. Slightly less obvious is how changes in the parameters of the distribution from which sampling takes place affect the equilibrium number of projects or sample size, N^*.

The two parameters of a normal distribution are the mean μ and the standard deviation σ. If x represents the random variable drawn from a standard normal distribution, then y is drawn from a general normal distribution if $y = \mu + \sigma.x$. Let v(N) represent the maximum of a sample of size N drawn from this distribution. The expected value of the maximum is therefore, on application to 2.2, approximately:

$$E(v(N)) = \mu + \sigma.a + \sigma.b.\sqrt{\ln N}.$$

Hence, as the mean of the distribution is increased, the expected benefit function in Figure 2.4 shifts vertically upward. Its effect is therefore similar to that of a decline in fixed costs – it affects the decision whether or not to sample in the first place, but does not influence the size of the sample N^* if any sampling at all is carried out. By contrast, the effect of an increase in the spread of the distribution, measured by σ, is to increase the slope of the benefit function for all N, and therefore to increase the equilibrium sample size N^*. This accords with intuition: more research is carried out the greater the uncertainty of the prospect, *ceteris paribus*.

2.4 SUPPLY OF RESEARCH OUTPUT

If the nature of inventive activity cannot itself provide a useful measure of research output, as section 2.2 above concluded, resort must be made to measuring research output by its effects. Research leads to technical progress, which can in principle be measured in the economic domains of prices, outputs, incomes, and so on. Ultimately, technical progress may be said to have occurred if given resources can achieve a higher level of aggregate consumer satisfaction than before. A definition along these lines encompasses the appearance of new products. However, abstracting from new product invention – or assuming that price indices are sufficiently sophisticated to allow for new products when they appear – technical progress might be measured by changes in total factor productivity or output per unit of input. Equivalently, it can be measured by the reduction in unit costs of output.

Most theoretical studies of invention assume a simple relationship between research output (unit cost reduction) and inputs of research resources. This might be called an 'invention production function', though William Nordhaus (1969) refers to it as an 'invention possibility function'.[16] Sometimes, as in Brian Wright (1983), the output of research is only specified to be an 'invention', and analysis proceeds in terms of its cost as a function of research inputs. Often it is assumed that the function is deterministic, but it always exhibits positive but diminishing marginal product; or in the case of a cost function, positive but increasing marginal costs. Where the function is stochastic, these features relate to the expected values of the respective variables.

In the paradigm of sampling from a normal distribution, presented above in section 2.3, the outcome was implicitly assumed to be a measure of value of the invention. This was adequate to

obtain the derived demand for sample size (research inputs), which
was identified with the expected marginal value product function,
equation 2.3. But, with the volume and unit value of research output undistinguished, it was not possible in that context to derive a
supply function, which of course expresses optimal ('volume') as a
function of its price (unit value).

The output of inventive activity is now assumed to be a reduction
in unit costs of production, labelled ξ. It is a function of research
inputs, Z:

$$\xi = f(Z), \text{ with } f(0) = 0, \text{ } f'(Z) > 0 \text{ and } f''(Z) < 0.$$

In a variation on this approach, the unit cost reduction ξ may be
for a particular invention and therefore given, at ξ^0 say, but
research resources Z may be applied to increase the probability of
actually achieving the invention, $P(Z)$. Hence the expected value of
the invention is $\xi^0.P(Z)$, and the ensuing argument in this section
may be interpreted in the context that $f(Z) = \xi^0.P(Z)$. This interpretation will be found useful later, in section 4.5, under the heading
of technological competition.

The economic value of the cost reduction is determined by the
volume of output to which it applies, Q, and research input is
chosen such that net value $\{Q.f(Z) - Z\}$ is maximised. Let Z^* be
the implicit solution to this maximisation; it solves $Q.f'(Z^*) = 1$.
It follows that Z^* varies directly with Q since
$dZ^*/dQ = -(Q^2f''(Z))^{-1} > 0$. Now, if the 'volume' of inventive
output is ξ and its value is $Q.\xi$, it follows that 'unit value' is Q.
Thus there is a curious duality in the inversion of the roles of price
and output between inventive and normal productive activity. The
volume of output to which a cost-reducing invention applies
represents the demand 'price' or unit value of the invention. Hence
the responsiveness to output of invention-induced cost-reduction is
the appropriate concept of elasticity of supply of invention.
Formally, this elasticity of supply may be defined as:

$$\eta = d\ln f(Z)/d\ln Q = [d\ln f(Z)/d\ln Z].[d\ln Z/d\ln Q]$$

where the first term of the multiplicative decomposition is the
elasticity of the invention possibility function, and the second term
is actually the negative of the inverse of the elasticity of the
'marginal invention possibility' schedule since the equilibrium
condition $Q.f'(Z) = 1$ implies that:

$$d\ln Z/d\ln Q = -d\ln Z/d\ln f'(Z)$$

Suppose, for example, that the invention possibility function has a constant elasticity, α, being:

$$f(Z) = AZ^\alpha$$

where A is a scaling constant. Then the 'marginal invention possibility schedule' is:

$$f'(Z) = \alpha AZ^{\alpha-1}$$

and the supply elasticity of inventions η is $\alpha/(1-\alpha)$. This one-to-one relation between the supply elasticity and the elasticity of the invention possibility schedule can be inverted as $\alpha = \eta/(1+\eta)$. The correspondence is of importance in assessing the empirical results reported in Chapter 9 below, where it is argued that the best estimate for η (when patent data are used as a proxy for ξ) is in the range $\frac{1}{4}$ to $\frac{1}{2}$, which implies a corresponding range of $\frac{1}{5}$ to $\frac{1}{3}$ for α. This is somewhat higher than the 0.1 assumed by Nordhaus on the basis of the empirical findings of Jora Minasian (1962; 1969) and Edwin Mansfield (1965), but Nordhaus reckons that his figure may be biased downwards because it makes no allowance for externalities. Moreover, it may not be unreasonable to assume that the elasticity of supply as estimated in Chapter 7 with patent data is higher than it would be were the output of research measured as the rate of unit cost reduction. The reason for this is explained in Chapter 7.

2.5 THE TIME–COST TRADEOFF IN RESEARCH

For many theoretical analyses of invention, especially where inventors are assumed to compete with each other as in Chapter 4 below, it is useful to cast the analysis in terms of the cost function for producing a particular invention. The important factor that this introduces is the idea of a trade-off between pecuniary costs and the gestation period of the invention. Thus it is generally assumed that the faster the research work is completed the higher the pecuniary costs of the research activity. This seems reasonable. But usually the further assumption that reduction in the gestation period involves increasing marginal costs is made. These ideas are now examined in the context of a simple model of research activity within the sampling paradigm.

The essential features of a time–cost trade-off can be derived from a sampling model of the research process in which the sampling may be carried out either sequentially or in parallel.[17] Sequential sampling takes longer to achieve any given result, but since on average fewer 'observations', or research projects, are needed, it is also cheaper than doing the projects at the same time, which is the parallel alternative. Actually, if there are more than two possible research projects, there is not just one alternative but a set of substitute mixtures of the sequential and the parallel approaches. This is what gives rise to the time–cost trade-off.

Assume that the research is oriented to a particular goal or set of goals, and that there are a number of different possible projects visible to the researchers at the outset. Assume further that each research project takes the same time, is independent of the others, and represents only a probability that the goal can be achieved.

An ordered sequence of projects will be called a research programme. It is possible for more than one project to coincide in the sequence, in which case the coincident projects are carried out concurrently and represent an element of parallel research. If all projects coincide in this way, then the research programme is one of pure parallel research. And if no projects coincide in the sequence, then the research programme consists of pure sequential research.

Consider how many possible research programmes exist when there are n research projects. The way to establish this is to find the total number of possible partitions of n entities. Let the number of ways in which a set of n elements can be partitioned into J parts $(0 \leq J \leq n)$ and permuted be represented by $\phi(n,J)$. This gives the number of different ways in which a research programme of n projects can be scheduled over J periods. To find the grand total of such possible paths as J ranges from 1, representing pure parallel research, to n, representing pure sequential research, $\phi(n,J)$ must be summed for all J up to n:

$$\Phi(n) = \sum_{J=1}^{n} \phi(n,J).$$

The fact that the number of research projects does not have to be large for the research administrator to have a very wide choice of ways of conducting the research programme, is illustrated in Table 2.1, which tabulates the functions $\phi(.,.)$ and $\Phi(.)$.

For a given programme of research projects, let p_i denote the

Table 2.1 *The number of potential research programmes*

Time span, J	$\phi(n, J)$ 1	2	3	4	5	$\Phi(n)$
No. of projects, n						
1	1					1
2	1	2				3
3	1	6	6			13
4	1	14	36	24		75
5	1	30	150	240	120	541

probability that at least one of the projects at the ith stage will be a success. Then, with a total of J stages, the expected time, measured in research project durations, for the research programme is:

$$E(T) = \sum_{j=1}^{J-1} j.(\prod_{i<j} (1 - p_i)).p_j + J.\prod_{i<J} (1 - p_i)$$

where the bracketed term within the summation denotes the probability of no success before the jth stage − and when j = 1 this term is taken to be unity.

Denoting by c_j the total cost of projects planned for the jth stage of the research programme, the expected cost of the whole programme is:

$$E(C) = \sum_{j=1}^{J} c_j.(\prod_{i<j} (1 - p_i))$$

The semi-ordered set of research projects which is organised into a research programme might have more or less projects running in parallel, and the combinations of expected time and cost corresponding to these possibilities imply an *ex-ante* trade-off between these overall characteristics of the research programme.

Two illustrations of the time-cost trade-off are now given. First, suppose that there are four research projects, each with probability of success equal to 0.5 and with cost equal to 1. Since the projects are similar in their basic characteristics, the only thing to distinguish one research programme from another is the number of projects carried out in parallel at any stage. This means, in effect, that the $\Phi(4) = 75$ potential research programmes boil down to only

Table 2.2: Expected time and cost for four similar projects

Programme	E(T)	E(C)
4	1	4
3,1	1.125	3.125
2,2	1.25	2.5
1,3	1.5	2.5
2,1,1	1.375	2.375
1,2,1	1.625	2.125
1,1,2	1.75	2
1,1,1,1	1.875	1.875

Notes $p_i = 0.5$ and $c_i = 1$ for $i = 1-4$.
Overall probability of success of each programme is $0.9375 = 1 - (1 - .5)^4$

eight distinct points in the $E(C) - E(T)$ space. The set of distinguishable combinations of expected cost and expected time is tabulated in Table 2.2 and displayed in Figure 2.5.

There are two points to note about this simple example. First, the 'output' of each and every research programme is identical, as it should be: in each the probability of overall success is 0.9375. Secondly, even in this simple case, it can be seen that the 'efficient

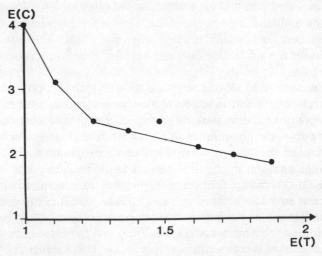

Figure 2.5: Time–cost trade-off for four similar projects

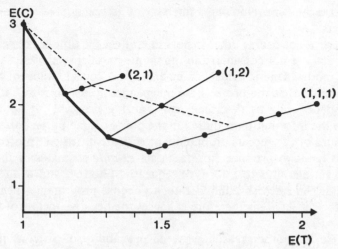

Figure 2.6: Time–cost trade-off for programmes of three distinct projects

Note: The projects have success probabilities of 0.3, 0.5 and 0.7, respectively; each absorbs one unit of costs and takes one unit of time to complete. Overall probability of success of each programme is 0.895.

set' of research programmes is a strict subset of all the possible programmes since one of the eight points is dominated by the others.

As a second example, Figure 2.6 shows the relationship between the expected cost and expected duration of programmes composed of three independent projects, each having a different probability of success, but with the same time and costs. The bold curve indicates the efficient set of programmes. Light lines connect programmes with similar stagings of the projects, which are indicated in brackets: thus (2,1) means that two projects are planned to be carried out in the first stage and one in a second stage. The dotted line connects a hypothetical 'average' set of programmes, constructed on the assumption that the research administrator cannot distinguish the quality of individual projects, but assigns a (correct) average probability of success equal to 0.5 to each.

What the example shown in Figure 2.6 makes clear, is that the efficient set has a marked downward slope and is convex; that the same is true of the average set of programmes, but in an attenuated form; and that there are considerable gains possible if projects do differ in success probability (or in cost) in ways that are recognised by the research administrator. This last point illustrates again the importance of preliminary filtering or screening of research projects.

There are of course other factors that affect the time–cost trade-off which are not considered in the simple parallel sequential sampling model. One is evident immediately on consideration of the assumption that the projects are independent. One form of interdependence may be described as a learning process – it applies where the information acquired in the doing of one project affects estimates of the success probability or cost of subsequent projects. If this tends to produce more accurate *ex-ante* estimates of these entities it amounts to a form of sequential filtering, and will tend to guide the research administrator to choose programmes nearer the efficient set. This therefore enhances the negative incline of the trade-off.

There is another reason why the probability of success of a particular project may depend on its position within the research programme. It arises from the likelihood that there are diminishing returns to increased research *within* a given time-period. Thus, as more projects are pursued concurrently, in parallel, lower-quality resources are devoted to the incremental projects. This amounts to saying that marginal costs are not constant with increasing intensity of research activity. This feature was not allowed for in the simple model where the cost of a research project was assumed to be the same whether it was carried out together with other projects or not. Since faster completion of a research programme implies doing more projects concurrently early in the programme, this factor should again increase the steepness of the time–cost trade-off.

Yet another, and quite separate, factor supporting both the negative incline and the convexity of the time–cost trade-off is the fact that when costs are being compared over time they should be made comparable by multiplying by a discount factor, such as e^{-rt} or $(1 + r)^{-t}$ where r is the rate of discount. Since the rate of change of the discount factor with respect to time t is negative but increasing, its imposition on the foregoing undiscounted cost–time trade-off again enhances the negative slope and convexity.

Parallel research efforts seem to be implicit in the 'crash programmes', which characterise spectacular R & D programmes such as the US Army's 'Manhattan Project' to build an atomic weapon, and the 'Apollo Project' which culminated in manned space flights to the moon. Both of these featured a 'race against time' element.

Indeed, much of military research activity may take this form, and it is noteworthy that a substantial part of the evidence on time–cost trade-offs is provided from this source. Among the important references are Frederick Scherer (1965; 1966), Merton Peck and Scherer (1962), A. W. Marshall and W. H. Meckling (1962), T. Marshak *et al.* (1967) and K. Hartley and W. Corcoran (1978). Taken together, these provide substantial empirical support for the relationship, but it is noteworthy how thin the documented evidence actually is in comparison to the importance it has assumed in theories of technological competition.

2.6 CONCLUSION

Although a number of approaches were adopted in this chapter, it has not been possible to arrive at any really strong predictions or conclusions. There is a suggestion, from the analysis that considered research as a form of sampling process, that there may be sharply diminishing returns to inventive activity. This notion is used extensively elsewhere in the theoretical sections of this book, and finds empirical support in Chapters 7 and 8 below. Another result of interest is the derivation of the commonly assumed time–cost trade-off from a simple model of substitution between sequential and parallel research activities in a research programme. This is made use of in the analysis of Chapter 5 below.

NOTES

1. The patent law definition can be found in Mansfield (1968), p. 50, for example. Siegel's definition is quoted in Kay (1979), p. 11.
2. The degree of belief interpretation of probability contrasts with the classical view of probability as the limiting value of an 'objective' relative frequency in repeatable experiments. Although it is now coming into fashion, especially in social and management sciences, it has as long a pedigree as the classical view, dating back at least to Jacob Bernouilli's *Ars Conjectandi* (1713). In more recent times the conversion from a strict relative frequency interpretation has been largely due to the philosophising of J. M. Keynes (1921), H. Jeffreys (1961) and F. P. Ramsey (1931).
3. Of course, it is possible to use inductive methods on problems of deduction, as is documented in Arthur Koestler's *Act of Creation*, where the evidence of the eminent mathematician Henri Poincaré,

among others, testifies to the power of 'unconscious mentation at
certain stages of creative work' (see Koestler, p. 166) which always
involves changing the given data of the problem with new evidence,
new analogies or a different viewpoint.

4. See for example, S. A. Schmitt: *Measuring Uncertainty* (1969), which
   contains a brief account of the entropy concept with suggestions as to
   possible uses. He concludes the section thus: 'For example, when
   several experiments are available you might choose to perform the one
   that is expected to reduce entropy the most' (p. 173).

   The use of the entropy of a distribution as a measure of uncertainty
   now has quite a large body of support in the literature, see in particular
   Tribus (1969), Cox (1961) and Khinchine (1957). For distributions
   defined over measured variables some measure of 'spread', such as
   variance or standard deviation or interfractile range may seem natural
   as a measure of uncertainty (although this is only intuitively so, because
   we tend to think in terms of centrally massed distributions). But there
   is no such 'natural' measure for distributions over categorical
   variables. In such cases what is looked for is rather a measure of un-
   evenness or roughness, which entropy serves well. In fact since it can
   be shown (see Nicoll, 1981) that scaling a continuous distribution by
   a constant factor will add the logarithm of that factor to the entropy
   of the distribution, just as it will change the standard deviation of the
   distribution by the same factor, it follows that there is a one-to-one
   correspondence between entropy and other measures of spread, and
   hence their interchangeability as far as measures of uncertainty are
   concerned.

5. These views of Schumpeter can be found, for example, in his *Business
   Cycles* (1939).

6. See, for example, his book (1966), or his co-authored articles with
   Z. Griliches (1963) and O. Brownlee (1962).

7. Of course, there are other, additional reasons why the industrial
   research laboratory might be a more efficient vehicle for the production of inventions than inventors working individually. Not least of
   these is the cost saving that can be achieved when several inventors
   share the same facilities, which the market may not be able to supply
   on a rental basis for the individual inventors.

8. This evidence is presented in some detail in Chapter 8 below.

9. A similar kind of distinction, derived from the demand side, is
   presented in Chapter 3. There the expressions 'minor' and 'drastic',
   which are standard in the literature, are employed. The terminology
   differs between Chapters 2 and 3 deliberately as a reminder of the
   different basis for the distinctions. However, there would no doubt be
   a correlation between the categories employed in the two chapters.

10. The main reference for evolutionary operations is the book by George
    Box and Norman Draper (1961), in Chapter 8 of which is given a brief
    bibliography listing over two dozen articles describing its operation in
    industry.

11. The stopping rule, of course, compares the costs and benefits of the last and next step in the sequential sampling process. The usual application of EVOP sees it as a part of a process monitoring and controlling a system which is in existence anyway, so this element of 'cost' can be ignored.

12. See, for example, Frederick Mosteller and Robert Rourke (1971, Chapter 15) for the general formula for the ith order statistic.

13. Stigler also noted the strong tendency to diminishing returns in the search for a minimum price: 'whatever the precise distribution of prices, it is certain that increased search will yield diminishing returns as measured by the expected reduction in the minimum asking price' (1961, p. 217). However, as Nordhaus (1969, 21) observes, the diminishing returns property of search does not hold for all distributions. A notable exception is the Pareto distribution with $\alpha < 2$.

14. It is pointed out in Mosteller and Rourke that $E(u)$ is linear in $\sqrt{\ln N}$. The coefficients of this relation used to draw the curve in Figure 2.3 were derived by least squares regression for $N = 5$ to 40, as $a = -0.783$ and $b = 1.5319$.

15. This formula was checked against the exact result from equation 2.1 and found to be accurate, despite the fact that its 'parent', equation 2.2, is an empirical approximation.

16. This function has also been posited by Edmund Kitch (1977), Pankaj Tandon (1982) and Partha Dasgupta and Joseph Stiglitz (1980), in all of which it is specified in the constant elasticity form when convenience dictates.

17. The first explicit recognition of the importance of considering parallel research in the economics of invention literature appears to have been by Richard Nelson (1961), whose main point is that parallel efforts should reduce the variance of the outcome within a given time. This is a different aspect of the matter from that of the text, though the validity of Nelson's proposition can be seen illustrated in Figure 2.6.

3 The Derived Demand for Inventions

This chapter examines the derived demand for invention. Assuming that an inventor can fully appropriate the value of his invention, the demand price of the invention is the same as the maximum income that he could receive for it, and can therefore be thought of as the incentive to invent. The starting-point for the analysis in this chapter is a well-known formulation of the problem in terms of the incentive to invent. This is presented in sections 3.2 and 3.3. However, the analysis of these introductory sections is incomplete because the cost function of the industry in which the invention will be applied ignores factor prices. This deficiency is corrected in section 3.4 which presents a modern formulation based on the concept of duality. The remaining three sections of the chapter explore the implications of relaxing some of the assumptions implicit in the basic model. First, the consequences of the temporary nature of patent protection are analysed in section 3.5. Then, in section 3.6, is an examination of the implications of a short-run cost function based on *ex-post* fixity of coefficients when there has been a history of technical change in the industry. Finally, section 3.7 presents an analysis of a revenue maximising monopolist's derived demand for invention.

3.1 INTRODUCTION

A distinction is often drawn between inventions of new products and inventions of new ways of producing existing products. The latter may be called cost-reducing inventions, and are much easier to analyse than new product inventions. However, the distinction may be less important than is sometimes believed. Inventions are valued not for their own sake but for the greater consumption possibilities they enable; so the value or demand for an invention, be it a new process or a new product, is a derived demand – derived from the conditions of supply and demand for ultimate consumption goods. Moreover, it has been argued (Lancaster, 1966) that many consumption goods are best thought of as joint intermediate products for more elementary consumption services, so the invention of a new consumption good can be thought of as a process invention which reduces the cost of particular combinations of those elementary consumption services, or consumption 'characteristics' as they are known. On these grounds it can be argued that all inventions fall into the cost-reducing category. Accordingly, in much of this chapter it will be assumed that the invention in question can be adequately described in this manner.[1] There are, however, certain kinds of product markets in which an essential feature of market behaviour is product differentiation. In these monopolistically competitive markets the derived demand for an invention may not be adequately captured by the pure cost-reducing paradigm.[2]

Inventions are by their very nature unique and indivisible. It is not possible to demand varying quantities of a given invention: the demand is for all or nothing, and the demand price for an individual invention is therefore a reservation price. This is a price that indicates the maximum that the buyer is willing to pay for the invention. Looked at the other way round, it is also the maximum gross income that the inventor can earn from the transaction. By summing all such potential transactions it is possible to derive the total amount which users are willing to pay for the invention. This is what Kenneth Arrow (1962) refers to as the incentive to invent. Arrow's analysis provides the starting point for the theory set out in this chapter.

Inventions as information or knowledge possess the public good characteristic of 'non-rivalry in consumption', which is to say that one person's or firm's use of the information cannot diminish its use elsewhere. This implies that the various distinct demands for an invention are correctly aggregated by vertical rather than horizontal summation of the individual demand curves. Of course, for an indivisible product the quantity axis of a demand curve degenerates to a single point and the demand price is a reservation price. The sum of all such reservation prices represents the derived demand for the invention or, *ex-ante*, the incentive to invent.[3] Perhaps it is best to call it the *potential* incentive to invent because another public good characteristic of information or knowledge is the high cost of exclusion in the absence of a well-policed property assignment. It is assumed in what follows that such a property assignment is indeed provided by a patent so that the potential incentive to invent is in fact appropriable.

3.2 THE DEMAND PRICE OF A PARTICULAR INVENTION

The basic analysis of this section establishes that the incentive to invent, or the maximum potential revenue from an invention, is a function of the importance of the invention; of its scale of application; of the elasticity of demand for the product to which the invention relates; and of the competitive structure and behaviour of the industry concerned. The partial equilibrium model set out below has been employed by a number of theorists in this area, beginning with Kenneth Arrow (1962) and J. S. McGee (1966).[4] The expression 'incentive to invent' is due to Arrow, and has been widely adopted by others subsequently, but in the present context it is useful to recall the equivalence of factor price with per unit factor income and to consider the invention itself as a factor of production, so that Arrow's incentive to invent can be renamed 'the demand price of the invention'. This leads naturally in section 3.4 to the derived demand for invention.

The concept of reservation price is perhaps normally thought of as a supply price for unique articles. For example, an object offered at an auction may only be sold as long as a certain price is exceeded. This minimum supply price is that object's reservation price. In like manner, the maximum price that a potential buyer is prepared to offer for an object is a reservation price on the demand side. It can be seen that when the good in question is the right to use an invention, the sum of all such reservation prices is precisely the maximum potential revenue from the invention or the incentive to invent. This is the demand price of the invention.

Consider now the long-run equilibrium position of the market for a product which can be produced with constant returns to scale in which price discrimination is not possible. With given cost and demand conditions the market may or may not be viable. If it is viable it may or may not be competitive. These possibilities are illustrated in Figure 3.1.

In Figure 3.1(a) the market is not viable because unit production costs exceed the demand price of output at all levels of output.

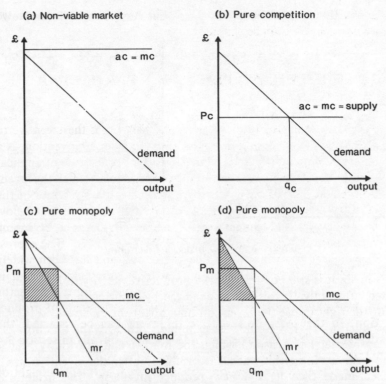

Figure 3.1: Market equilibria under competition and monopoly

Figure 3.1(b) illustrates the case of pure competition in which the supply curve is identical to the constant long-run marginal cost curve. Figures 3.1(c) and 3.1(d) illustrate the pure monopoly case, with equilibrium output determined by inter-section of the marginal revenue and marginal cost curves; monopoly profit is given by the hatched area in these diagrams, which are, of course, equal.[5]

An invention now appears which has the affect of reducing the long-run marginal costs of production. Suppose that the inventor has a watertight patent on his invention which he can license out on his own terms. What is the maximum reward he can extract from this property? A pictorial answer to this question is provided in Figures 3.2 and 3.3. Whatever the original form of the market, the patent ensures that its essential feature is now monopolistic.

(a) One representation **(b) Alternative representation**

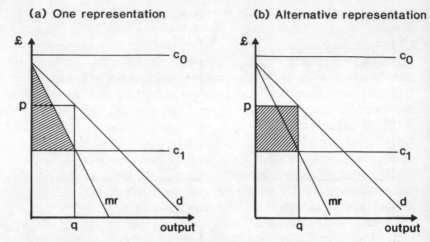

Figure 3.2: New product invention

Even so, it is still possible for the productive activity to be organised competitively; this can be achieved by ensuring that payments for the right to use the invention only affect the marginal costs of production, as they would, for example, with a royalty on output. Consider first the case in which the market had previously been non-viable. If anything at all is produced after the invention has been made then it is a new product invention. The inventor

(a) Minor invention **(b) "Drastic" invention**

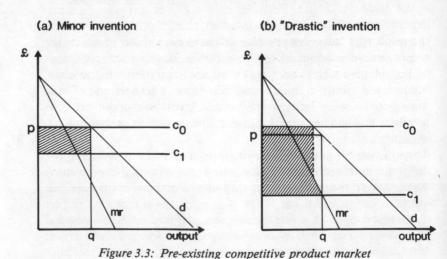

Figure 3.3: Pre-existing competitive product market

could license out the invention to a newly-created competitive industry, setting a revenue-maximising royalty per unit of output. Equally, he could auction off the right to use the invention to the highest bidder, thereby creating a zero-profit production monopoly. Or conceivably, he could set up in production himself. Analytically, all three possibilities are equivalent and are depicted in Figure 3.2 where c_0 represents the pre-invention level of unit costs and c_1 is the level of unit costs excluding any royalty element after the invention has been made. The per unit royalty is, of course, the difference between the price p and the post-invention level of unit costs. Revenue from the invention, equal to the area between the marginal revenue curve and the new marginal cost curve, which is shaded in the diagram, is also expressed as $q.(p - c_1)$.

Turning now to the case in which a competitive market existed before the advent of a cost-reducing invention, the same three possibilities confront the revenue-maximising inventor and again they are analytically equivalent. There is however a new twist to this story, which arises from the fact that it is still possible to produce with the previous technology at the pre-invention level of unit costs. Thus price is constrained by competition to lie below this level. This means that the effective residual demand for output under the new technology is traced out by the lower of two curves: the one being the post-invention unit cost curve, and the other being the original demand curve. With this adjustment the analysis is the same as before. But it implies that the effective residual demand curve is kinked at the intersection of the two curves, with the result that the corresponding effective marginal revenue curve is discontinuous at that output level. This is shown in the two parts of Figure 3.3 which illustrate respectively the distinction between 'minor' and 'drastic' inventions.[6] The distinguishing feature of a drastic invention is that it results in a lower selling price of the product; with a minor invention the selling price is unchanged. It is straightforward to show[7] that for any downward-sloping demand curve the dividing-line between a minor and a drastic invention is an invention which leads to a proportionate reduction in costs equal to the inverse of the elasticity of demand at the original output.

It may be noted that for competition to remain viable in the post-invention product market, payment to the inventor must be in the form of a royalty per unit of output sold. If the right to produce with an extramural invention were secured by a lump-sum payment then unit total costs would be a declining function of output for the licensee and a monopolistic market would be brought into being. This would not affect the maximum returns to the inventor since the licensee will produce at that output for which the new level of marginal cost equals the marginal revenue implicit in the derived demand for output under the new technology; and competition among potential licensees will ensure that all the monopoly profit goes to the inventor.

Finally, the case of monopoly in the product market must be examined. This case is more ambiguous than those considered so far because it hinges on the nature of the monopoly. It has already been pointed out[8] that some sort of entry barrier must be invoked to justify the monopoly assumption since it does not follow from the assumed cost function. It is natural to ask whether, and to what extent, there might be entry barriers after the invention has been made. In fact, this is the critical question when it comes to determining the maximum possible revenue the inventor can extract from the invention. The two extremes are illustrated in Figure 3.4. Part (a) assumes that post-invention entry barriers remain insurmountable, whereas in part (b) they are assumed away entirely. Obviously the latter case is identical to the foregoing competitive product market, though only a minor invention has been illustrated in Figure 3.4. Previous analysts seem to have ignored this case, assuming instead that the pre-existing level of monopoly profit is maintained,[9] which is the situation in part (a). However, comparing the two parts of the figure, it is clear that for part (a) to be applicable, as has hitherto been assumed by other writers, the post-invention barriers to entry must be at least as large as the area of the quadrilateral which is the difference between the respective shaded areas in the two parts of the diagram. This is the less plausible the larger the invention. Of course the barriers to entry may be statutory rather than financial, in which case the same quadrilateral area measures the amount by which they diminish the incentive to invent.

Assume now that there are large post-invention barriers to entry in this industry so that part (a) applies. Whether the whole of the shaded area accrues as revenue to the inventor depends first, on whether the invention is intra- or extramural, and then, if the latter, on the bargaining strengths of the respective bilateral monopolists.

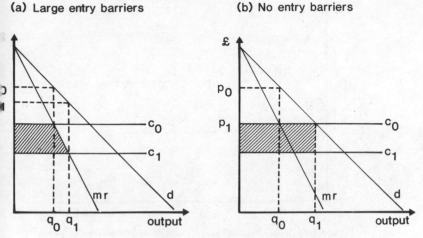

(a) Large entry barriers   **(b)** No entry barriers

Figure 3.4: Pre-existing monopolistic product market

An intra-mural invention is one produced by the production monopolist himself, in which case the whole of the shaded area in part (a) accrues to him and represents his incentive to invent. The bilateral monopoly situation that emerges from extramural invention is examined in the following section.

3.3 EXTRAMURAL INVENTION WITH BARRIERS TO ENTRY

When the product market is monopolistically organised with barriers to entry, the emergence of an invention for which the inventor is protected by an exclusive patent gives rise to a case of bilateral monopoly. Usually in cases of bilateral monopoly there is no clear solution for the price of the commodity in question since one party gains at the expense of the other, and particular assumptions about bargaining strategies are needed to resolve the matter. In the case at hand, what is being traded is the right to use the invention in production. It turns out that some definite predictions are possible, without resort to particular and arbitrary assumptions about bargaining. It will be shown first, that the incentive to generate a given cost-reducing invention extramurally must be less than it is for the product monopolist to do so intramurally; and secondly,

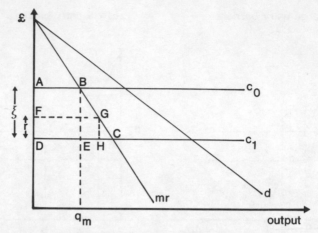

Figure 3.5: Monopoly product market with barriers to entry

that payment to an extramural inventor should take the form of a lump sum unrelated to output rather than a royalty on output. [10]

Consider a given cost-reducing invention which reduces unit costs from c_0 to c_1. Let $c_0 - c_1 = \xi$. Assuming the inventor prices his invention on a royalty per unit of output basis, consider the gains to the inventor, G^i, as a function of the royalty rate r. With $r = \xi$, $G^i(\xi)$ is given by the area ABED in Figure 3.5, since output remains at the intitial level q_m as the producer's marginal costs stay at the pre-invention level.

An $r > \xi$ is, of course unfeasible. For $0 < r < \xi$ such as that illustrated in Figure 3.5, G^i is given by the area of the rectangle DFGH. As r is increased from 0 to ξ it can be seen that G^i increases from zero and reaches a maximum either at $r = \xi$ or at $r < \xi$ at a point given by the condition that the absolute output elasticity of the function 'marginal revenue minus post-invention costs' equals one. An illustrative G^i for this case is sketched in Figure 3.6.

Now consider the gains accruing to the monopoly producer, $G^m(r)$, as the royalty rate is varied. It can be seen that $G^m(\xi)$ is zero and $G^m(0) = \max G^m(r)$ is given by the area ABCD in Figure 3.5. For a value of r lying between 0 and ξ as illustrated in that diagram, $G^m(r)$ equals the area ABGF, and the derivative $G^{m\prime}(r)$ is given by the horizontal distance FG. The function $G^m(r)$ is therefore downward sloping and convex. The two curves $G^i(r)$ and $G^m(r)$ and their sum are drawn in Figure 3.7.

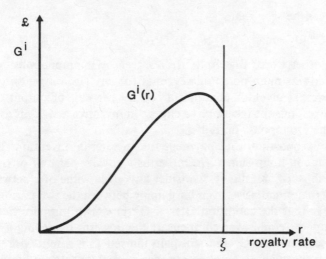

Figure 3.6: Inventor's revenue function

Considering the sum $G^i + G^m$, it can be seen that it is less than $G^m(0)$ by the amount given by an area such as GHC in Figure 3.5, and it is evident that this amount increases at an increasing rate ($= r$) as r increases.

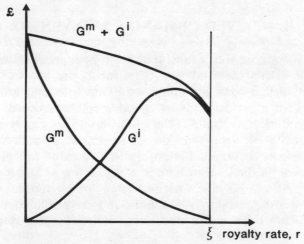

Figure 3.7: Inventor's and monopolist's revenue functions

It can be seen that:

$$G^m(0) = \max G^m(r) > \max G^i(r)$$

or in other words that in-house research by the monopolist yields greater maximum potential revenues for any given invention than an external inventor charging a royalty per unit of output could achieve. Thus the incentive to engage in inventive activity is greater for the monopolist himself.

The proposition that payment for an external invention should be made in lump-sum form rather as a royalty per unit of output is established as follows. Consider any given value of r between 0 and ξ that might arise from bargaining between the two parties and observe that the condition $G^i(r) + G^m(r) < G^m(0)$ implies that the monopolist could over-compensate the inventor by paying a lump sum greater then G^i and still gain himself as a result. Moreover, this arrangement to compensate the external inventor in lump-sum form is not only predictable, it is also superior on welfare grounds. The lump-sum payment leaves the monopoly producer's post-invention marginal cost curve unaffected, unlike the per unit royalty arrangement which raises it *pro rata*, so final output of the product is higher and its price lower. There is accordingly a larger consumer surplus under the lump-sum arrangement than under a per unit royalty scheme.

3.4   THE DERIVED DEMAND FOR INVENTIONS

For an industry in which firms treat their input prices parametrically, the derived demands for those inputs are implicit in the assumption that costs are minimised. That is to say, firms formulate their input demands by choosing combinations of inputs with given prices to minimise the cost of producing any given output. Thus the demand function for any input is implicit in the assumption of cost-minimisation, and such demand functions are conditional on the level of output as well as on all factor prices. Obviously they also reflect the technology of production; indeed both the cost functions of the firms and the conditional factor demand functions imply and are implied by the respective production functions.

It is conventional to define both the production function and its implied mimimum cost function for a given state of technology. Thus the firm's problem is formulated as:

$$\text{minimise } C = \sum_i w_i x_i, \qquad i = 1, \ldots, n \qquad\qquad 3.1$$

such that $q = f(x)$ where $x = \{x_1, \ldots, x_n\}$.

Here C represents total cost, w_i the unit price of the ith factor, x_i the quantity of the ith factor, and q the predetermined level of output. The solution to this constrained optimisation problem is a set of optimal values for the x_i's. Denote these by x_i^*. They are each functions of the n factor prices w_i and of output:

$$x_i^* = x_i^*(w_1, \ldots, w_n, q). \qquad\qquad 3.2$$

Substituting these optimal values into the objective function of the problem 3.1 yields minimum cost C^* as a function of the same set of variables:

$$C^* = C^*(w_1, \ldots, w_n, q). \qquad\qquad 3.3$$

Now, by Shephard's lemma of the envelope theorem,[11] the conditional factor demand functions 3.2 are simply the n partial derivatives of the mimimum cost function with respect to the factor price in question.

$$x^*(w_1, \ldots, w_n, q) \equiv \partial C^* / \partial w_i$$

These factor demand equations are conditional on the level of output.[12]

Now consider the technology itself to be an input. That is, suppose that the firms using the technology need to make payments to a patent holder who has a property right in the intangible property that the technology represents. It may be possible for these firms to choose an alternative technology, but they must opt for one or the other. Their decision is 'all or nothing'; they cannot in the nature of things consider mixtures of technologies. It is true of course that there are enterprises which produce with more than one alternative technologies – for example, a steel producer may use both the old Bessemer and the new oxygen processes in different plants. But if the productive activities of such enterprises are separable, as is likely, they may just as well be thought of as two distinct firms. The all-or-nothing nature of technology as an input can be represented by a switching variable – one that takes the value 1 if the technology is used, and 0 if it is not used. Let z_j represent such a variable for the technology indexed by j. Then the production function might be represented as:

$$q = f(x_1, \ldots, x_n; z_1, \ldots, z_m) \qquad 3.4$$

where there are n conventional factors and m technologies available.

For example, suppose that two distinct technologies are available to a producer. For concreteness let the first be an isoelasticity Cobb–Douglas function, $q = (x_1.x_2)^{1/2}$, and let the second be a fixed coefficient Leontief function, $q = \min\{x_1, x_2\}$. Overall, the production function can be written:

$$q = z_1.(x_1.x_2)^{1/2} + z_2.\min\{x_1, x_2\}. \qquad 3.5$$

The *ex-ante* unit isoquant, unit isocost and unit factor demand functions for this composite technology are shown in Figure 3.8 on the assumption that both technologies are freely available. Neither technology dominates the other for all factor price ratios: in fact the Leontief is chosen if the factor price ratio w_1/w_2 lies between α_2 and α_1 on the figure, and the Cobb–Douglas technology is chosen for factor price ratios outside this range. Since both technologies are available *ex-ante*, it is relevant to consider the convexification of the overlaid unit isoquants in part (a) of Figure 3.8 as the overall isoquant for the composite of possible factor combinations that these two technologies offer. The cost and factor demand functions in parts (b), (c) and (d) of the figure relate to this convexification.[13] In this example both production functions are homogeneous of degree one and hence possess the homotheticity property that the slope of isoquants is constant on radial expansion, implying that only one isoquant completely characterises the technology. This means that the level of output is immaterial regarding the minimum cost preference of one technology over the other. Note, however, that this is not true in general, though it will continue to be assumed here in line with the assumption of constant returns to scale of section 3.3.

It is useful to expose the construction underlying parts (c) and (d) of Figure 3.8. This is done in Figure 3.9. The overall cost function for the composite technology implied in the production function 3.5 is the minimum of the two separate cost functions implicit in

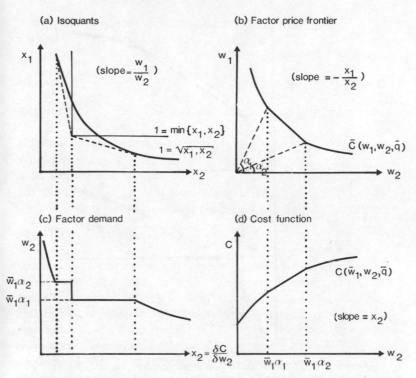

Figure 3.8: Substitute technologies: production, cost and factor demands

the respective components of the production function:

$$C(w_1, w_2, q) = \min \{ C^1(w_1, w_2, q), C^2(w_1, w_2, q) \}$$

where the superscript [1] refers to the Cobb–Douglas component of technology and superscript [2] refers to the Leontief component. The latter is linear in factor prices as substitution possibilities are absent.

Suppose now that one of the component technologies of the composite production function 3.4 or 3.5 is freely available and that the other relates to a new invention. With output and factor prices given, the social value of the invention is the reduction in total industry cost compared to the pre-existing technology. Assuming that this social value can be fully appropriated by the inventor, it represents the demand price of the invention. For linear homogeneous production functions the curves in Figure 3.9 apply

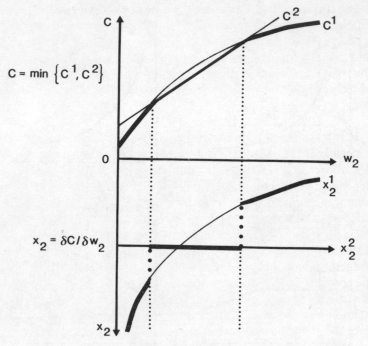

Figure 3.9: Cost and factor demand for composite technology

to the industry, and the reduction in industry cost can be read off the upper section of the diagram as the vertical difference between the cost curves. If technology 1 is pre-existing and technology 2 relates to the new invention, the price of which is v_2^*, then

$$v_2^* = \begin{pmatrix} C^1 - C^2 \text{ for } C^2 < C^1 \\ 0 \qquad \text{for } C^2 \geqq C^1 \end{pmatrix}$$

It is now obvious that the demand price of the invention is not only a function of the level of output, as was shown in section 3.3, but is also a function of the prices of the standard factors of production:

$$v_2^* = v_2^*(w_1, w_2, q)$$

In general the value of the invention of technology 2 when technology 1 is freely available is given by:

$$v_2^* = \max \ \{(C^1(w_1, w_2, q) - C^2(w_1, w_2, q), 0\} \qquad\qquad 3.6$$

But for linear homogeneous production functions, the conditional total cost function $C(w_1, \ldots, w_n, q)$ can be written $q.c(w_1, \ldots, w_n)$, where $c(w_1, \ldots, w_n)$ is the average cost function and possesses the same homogeneity and concavity properties as the total cost function. Hence the demand price for the invention of technology 2 when both are linear homogeneous can be written:

$$v_2^* = q.\max \{ (c^1(w_1, w_2) - c^2(w_1, w_2)), 0 \} \qquad 3.7$$

It is the level of output multiplied by the change in unit costs, if negative. This of course is in line with the exposition of section 3.3 above; the difference is that equation 3.7 emphasises the dependence on factor prices. Since the unit cost functions of 3.7 are homogeneous of degree one in factor prices, so too is the demand price of the invention.

One feature that stands out clearly on consideration of Figure 3.8 is the fact that an invention may be economically viable at one set of factor prices but unviable at another. Only inventions that lower the cost curve for all factor prices are unequivocally superior to the pre-existing technology. Even for such unequivocally superior technologies the demand price, and hence the inducement to produce the invention, will in general vary with factor prices. It is only in the case of a parallel downward shift in the cost curve that the demand price of the invention is independent of factor prices – and such an invention may be characterised as one which leaves all factor demand schedules unaffected (because the slope of the cost function represents factor demand by Shephard's lemma). But now, if factor prices are given and conditional factor demands stay constant, there can be no decline in total cost unless cost is not equal to the sum of all factor payments. But this is only possible if there is some fixed element in costs, with the invention only having the effect of reducing the cost of this fixed component. It must therefore be concluded that for long-run cost curves the demand price of an invention cannot be independent of factor prices.

That the economic importance, and even relevance, of an invention depends on factor prices has a number of implications. One of these relates to the notion of 'appropriate technology', by which is usually meant the fact that international technology transfer from advanced to technologically backward societies may not be very beneficial. A justification for this can be founded on different relative scarcities, and hence prices, of factors of production. For the same reason it can be expected that the new technologies actually adopted and invented in countries with different relative factor prices will reflect these differences. Even for a particular country over long periods of time, when relative factor prices may diverge from their intitial configuration, if might be expected that the characteristics of inventions change as a consequence of the changing demand prices of different types of inventions reflecting the changing relative factor scarcity.

An upshot of the dependence of the demand price of inventions on relative factor prices is the theory of induced bias in technical change. This is discussed in more detail in Chapter 4. The basic idea, however, is due to Sir John Hicks (1932), who argued that changing relative factor prices will tend to induce inventions that represent technologies that are less intensive in the factors of production for which relative price has increased.

It was suggested at the beginning of this section that an invention of a new technology might itself be considered as a factor of production, supplied in an 'all or nothing' manner. But so far the analysis in terms of optimal cost and factor demand functions has not involved the price to the producing units of using the invention. Instead the derived demand price of an invention has been implicitly formulated in a residual manner, like a rent. No doubt this is the correct way of modelling it at the level of the industry. But for a firm that is small in relation to the industry, the price it must pay to use a patented technology should be treated exogenously. It is a facet of the environment in which the firm is operating. Let the price of technology i be labelled v_i. Then the cost function represented in equation 3.3 should be amended to:

$$C^* = C^*(w_1, \ldots, w_n, v_1, \ldots, v_m, q) \qquad 3.8$$

Suppose that technology i is actually in use, and consider how the firm's costs given by 3.8 are affected by an increment in the price of technology. The technology either continues in use at the new price, or else is completely substituted by an alternative technology. In the former case the increment in costs must be equal to the price increment, while in the latter case the increment in cost is the difference between the old costs, including the price of the old technology, and the costs of producing the given output with the substitute technology. But, considering infinitessimally small increments in the price of technology i, the critical point at which production switches to the substitute technology is characterised by equal total cost of production under the alternative technologies. This implies, therefore, that:

$$\partial C^*/\partial v_i = 1 \text{ for } v_i < v_i^*$$
$$= 0 \qquad v_i \geqq v_i^*$$

where v_i^* represents the critical (limit) price of technology i, at which it is perfectly substitutable by the alternative technology.

The kink in the cost function at $v_i = v_i^*$ implies that it is no longer differentiable at that point. This signifies that the derived factor demand function for the technology has a jump at that price, which is of course to be expected from the all-or-nothing nature of technology as a factor input. Indeed it is an expression of the fact that demand for a technology must be modelled as a reservation price; Figure 3.10 illustrates.

The determinants of the reservation demand price v_i^* were outlined earlier in equation 3.7 for a unit cost reduction with con-

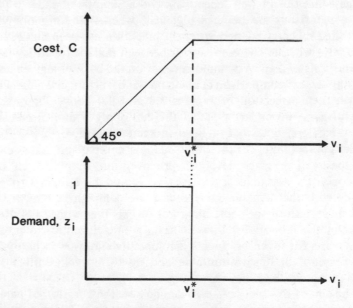

Figure 3.10: Cost as a function of the price of technology i and implied demand for that technology

stant returns to scale technologies. Thus

$$v_i^* = v_i^*(w, q) = q.\max\{c^o(w) - c^i(w), 0\}$$

where the unit cost function $c^i(w)$ relates to the technology in question and $c^o(w)$ is that for the best alternative technology at factor prices w. Note that, for constant returns to scale technologies, v_i^* is proportional in output q. This is the basic justification for using output as a proxy for demand price in the empirical analyses based on Schmookler's (1966) patent data reported in Chapter 6, in which it is argued that the statistical relation between numbers of invention and output identifies a supply function of inventions.

3.5 DURATION OF PATENT PROTECTION

An important motivation of the literature on the incentive to invent has been the question whether monopoly in the product market provides greater incentives for invention than does a competitively organised product market. The foregoing sections of this chapter suggest that the answer depends on two assumptions under which the comparisons are being made: first, whether the monopoly is protected by entry barriers *after* the invention has been made: and secondly, whether the comparison between market forms is made on the basis of a given demand curve or on the basis of a given level of output. [14] If comparison is made on the basis of a given demand curve, then inspection of Figures 3.3 and 3.4 above shows that incentives to invent are equal if the monopoly is unprotected by entry barriers, but when it is so protected incentives are greater under competition. On the other hand, if comparison is made on the basis of a given level of pre-invention output, then the monopolist's marginal revenue curve should be adjusted to coincide with the demand curve facing the competitive industry, [15] and inspection of the same diagrams reveals that without barriers to entry the monopolist himself has a greater incentive to invent than exists for inventors facing a competitive industry, whereas in the presence of unsurmountable barriers to entry incentives are approximately the same. [16]

The aim of this section is to examine how the duration of patent protection affects a comparison of the incentive to invent between the polar cases of competition and monopoly in the product market. It is assumed that the comparison is on the basis of a given demand curve for the product, and that where the market is organised as a monopoly, it is protected by entry barriers of some kind. With a given demand curve for the final product and insuperable barriers to entry protecting a monopolist, the derived demand for invention is greater when production is organised competitively because of the normal output-limiting effect of monopoly. This is obviously true if the duration of patent protection exceeds the expected economic life of the invention. But it is not necessarily true if patent protection ceases before the invention is obsolete. Suppose, to begin with, that the expected economic life of the invention – the period within which a 'replacement invention' is not expected to materialise – is infinite, but that patent protection on that invention is limited to T years. Clearly it is possible to imagine a patent period short enough to reverse the conclusion that competition offers greater incentives to invent than monopoly for any 'size' of invention since in the limit as the period of patent protection goes to zero so too does the discounted value of revenue from the invention with a competitive product market for all sizes of invention.

Denote by $R^c(\xi)$ the annual rate of revenue generation from an invention that reduces unit costs by ξ when the product market is competitive, and by $R^m(\xi)$ the corresponding function for the monopolist inventor. These revenue streams are assumed to last for ever, and the functions are depicted in Figure 3.11, which illustrates Arrow's contention that competition offers greater incentives to engage in inventive activity than does monopoly. But now, if the period of patent protection is different from the length of time that the monopolist can enjoy his revenue stream, a comparison requires that the two functions must be suitably discounted.

It is straightforward to show that the critical patent period that is short enough to reverse Arrow's conclusion is given by a simple function of the rate of discount, ρ, and the ratio of competitive to monopolistic output at the initial level of unit costs, ν. Define

$$D^c(\rho, T, \xi) = \int_0^T R^c(\xi).e^{-\rho t}\, dt = (1 - e^{-\rho t}).R^c(\xi)/\rho$$

so $\quad D^c(\rho, \infty, \xi) = R^c(\xi)/\rho$

and $\quad D^c(\rho, T, \xi) = (1 - e^{-\rho T}).D^c(\rho, \infty, \xi)$

and $\quad D^c_\xi(\rho, T, \xi) = (1 - e^{-\rho T}).D^c_\xi(\rho, \infty, \xi)$

$$\qquad\qquad = (1 - e^{-\rho T}).R^c_\xi(\xi) \qquad\qquad 3.9$$

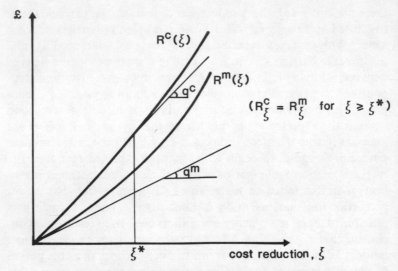

Figure 3.11: The functions $R^c(\xi)$ and $R^m(\xi)$

where $D\,(\rho, T, \xi)$ represents the discounted timestream of revenue
from an invention which reduces unit costs by ξ, and which is sold
to a competitive industry until expiry of the patent in T year's time.
The superscript ^{c} denotes competition, and subscripts denote
partial derivatives.

For the monopoly case, in which the economic life of the invention is assumed infinite, the discounted time stream of revenue is:

$$D^m(\rho, \xi) \equiv \int_0^\infty R^m(\xi).e^{-\rho T}\,dt = R^m(\xi)/\rho$$

and

$$D^m_\xi(\rho, \xi) = R^m_\xi(\,\xi)/\rho \qquad\qquad 3.10$$

where the superscript ^{m} denotes monopoly. Figure 3.12 exhibits the
analysis in terms of present values.

Now define $\nu \equiv Q^c/Q^m$, the pre-invention ratio of output under
competition and monopoly. Arrow (1962, footnote 8) has shown
that for very small inventions the ratio of incentives to invent under
competition and monopoly is approximated by ν. This can be seen
by comparing figures 3.3 and 3.4. It is clear therefore that:

$$R^c_\xi(0) = \nu.R^m_\xi(0)$$

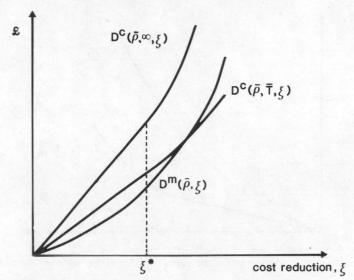

Figure 3.12: Discounted revenue functions for competition and monopoly

Hence, from equations 3.9 and 3.10:

$$D^c_\xi(\rho, T^*, 0) = \nu.(1 - e^{-\rho T^*}).D^m_\xi(\rho, 0)$$

and

$$D^c_\xi(\rho, T^*, 0) < D^m_\xi(\rho, 0) \text{ implies } \nu.(1 - e^{-\rho T^*}) < 1$$

which in turn implies that

$$T < -\ln((\nu - 1)/\nu)/\rho = T^*, \qquad\qquad 3.11$$

where T^* is the longest period of patent protection for which the incentive to invent is greater under monopoly than under competition. The effect of a shorter patent period, T, is to pull down the curve $D^c(\rho, T, \xi)$ and to give it a smaller slope in the ξ direction for any given value of ξ. The general shape of the curve, linear up to ξ^* and thereafter convex, is retained.

Since the D^m curve is convex throughout its length, the critical value of T such that incentives are greater under monopoly for all values of ξ is given by the condition that $D^m_\xi(\rho, 0) = D^c_\xi(\rho, T^*, 0)$. Figure 3.13 illustrates the situation.

Inserting plausible value for ρ and ν, suppose that $\nu = 2$, which would apply for a linear demand curve, and that $\rho = 0.05$. Then

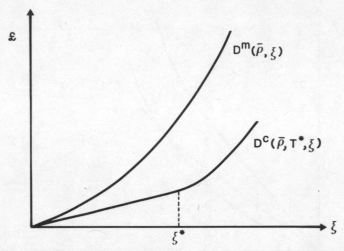

Figure 3.13: Incentives under monopoly exceed those under competition

inequality 3.11 gives:

$$T < T^* = 13.9 \text{ years}$$

which is remarkably close to the typical maximum patent duration in effect nowadays. Now, Arrow (1962) claimed that 'the only ground for arguing that monopoly may create superior incentives to invent is that appropriability may be greater under monopoly than under competition'. But here is a case where appropriability *is* indeed greater under monopoly because of the finite patent period, and to such an extent that Arrow's conclusion is reversed for almost any size of invention.

Suppose now that the duration of patent protection is greater than T^* for particular values of ρ and ν; it may then happen that competition provides superior incentives for small inventions whereas monopoly gives superior incentives for larger inventions as figure 3.14. illustrates.

So far it has been assumed that the expected economic life of the invention is infinite, or that its embodied technology is not expected to become obsolete in the future. This is clearly unrealistic. Suppose instead that the invention's expected economic life is limited to L years. Now the discounted revenue streams due to the invention under monopoly and competition are:

$$D^m(\rho, L, \xi) = \int_0^L R^m(\xi) . e^{-\rho t} dt$$

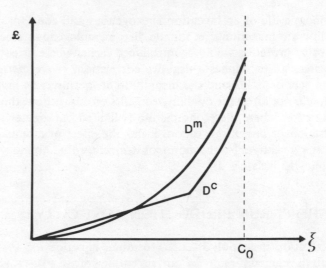

Figure 3.14: Reversal of superior incentives with size of invention

and

$$D^c(\rho, L, T, \xi) = \int_0^{\min(L, T)} R^c(\xi).e^{-\rho t}\, dt$$

respectively. The qualitative conclusion that competition in the product market offers greater incentives to invent than monopoly remains intact if the economic life is shorter than the duration of patent protection, i.e. if $L \leq T$. When $L > T$, it can be shown[17] that the duration of patent protection short enough to reverse that qualitative conclusion for all sizes of invention is such that the following inequality must be satisfied:

$$T \leq -\ln\{(v - 1 + e^{-\rho L})/v\}/\rho \qquad\qquad 3.12$$

Again, inserting plausible values for the parameters, let $v = 2$, $\rho = 0.05$ and $L = 20$ years. Then the condition 3.12 implies that

$T \leq 7.6$ years.

What this section has shown is that, notwithstanding the standard static analyses comparing incentives to invent under monopoly and competition, recognition of the temporary nature of patent protection implies that there are no simple, clearcut propositions to suggest that one form of market organisation is superior to the other on this score. In making this point it has been assumed that the economic life of an invention is exogenous and constant. But in reality an invention's economic life is endogenous since a superseding invention would be introduced earlier where incentives are greater, which implies a negative correlation, *ceteris paribus*, between economic life and the magnitude of incentives to invent. This should not affect the qualitative results of this section, though in those cases where the economic life is limited but greater than the duration of patent protection, it has the effect of attenuating the greater incentives for the monopolist-inventor that obtain when equation 3.12 is satisfied.

3.6 SHORT-RUN PRODUCTION COST CURVES

The derivation in section 3.4 of a formula, equation 3.6 for the reservation demand price of an invention of a cost-reducing technology applies to both the long and the short run. That is to say, it requires only the appropriate modification of the relevant cost functions to allow some fixity in inputs for the general formula to apply to the short run. But whether some factor or factors should be considered fixed, and if so which ones, for the determination of the reservation demand price of an invention is an empirical matter. It depends on both the characteristics of the invention and the span of time for the short-run/long-run distinction compared to that required to appropriate the benefits of the new technology. Obviously the duration of patent protection, discussed in section 3.5, is relevant for the latter consideration.

But while it may be realistic to assume some factors of production to be fixed, there is unfortunately little that can be said about the consequent demand for invention without some further assumptions. Its social value is given by the reduction it allows in industry costs, which is implied by equation 3.6, but to go beyond this some specific assumptions about the nature of short-run costs are needed. As an illustration, the vintage production function model introduced independently by W. E. G. Salter (1960) and Leif Johansen (1959) is used in this section as a particularisation of production and cost functions that allows the short run to be analysed. The exposition is entirely in diagrammatic terms. The basic model is set out initially, after which the incentives to engage in inventive activity under both competition and monopoly in the product market are examined.

Any firm is assumed able to produce current output either with installed machines which are its heritage from past investment, or with new equipment. The installed machines are the 'fixed' factor which transforms the analysis from the long run to the short run.

The set of installed machines can be partitioned into subsets corresponding to their various vintages. The unit operating costs of machines of any given vintage are identical and constant but they may vary between vintages. The profit-maximising firm will employ all machines for which unit operating costs are less than the price of the product (under competition) or marginal revenue (under monopoly). All other machines are obsolete. Thus the marginal vintage of machines is that with the highest unit operating costs less than or equal to price or marginal revenue. If the various vintages are arranged in ascending order of their unit operating costs, which may, but need not, correspond to their chronological order by vintage, with total output measured on the horizontal axis and unit operating cost on the vertical axis, the short-run marginal cost curve for any firm is arrived at. It assumes that only installed equipment is available, and is pictured in Figure 3.15.

The firm has a choice between producing with its inherited machinery or with new equipment. With new equipment all factors are variable, thus the relevant cost curve is the long-run cost curve,

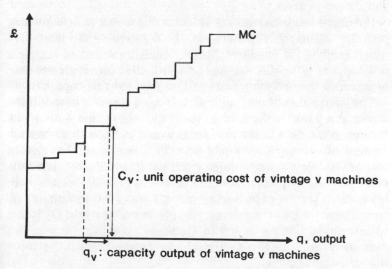

Figure 3.15: The short-run vintage marginal cost curve

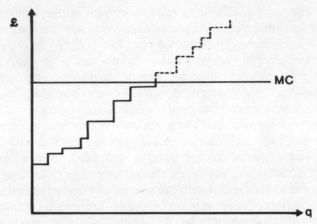

Figure 3.16: Long- and short-run marginal costs

assumed to exhibit constant average and marginal costs. This curve may be drawn in to show the overall marginal cost curve of the firm, which at any output rate will be the lower of the fixed and variable equipment curves. This is illustrated in Figure 3.16.

For a competitive industry these marginal cost curves may be summed horizontally to give the industry supply curve. Equilibrium price and output are shown by the intersection of the supply and demand curves.

If demand has been constant or increasing over a sufficiently long period, [18] a 'snapshot' picture may be depicted as in Figure 3.17 which exhibits the conditions: price equals total cost of marginal new capacity (long-run marginal cost), which in turn is greater than or equal to the operating costs of marginal existing capacity.

The intersection of the supply and demand curves is purposefully drawn at a 'kink' in the former, for if the intersection were drawn further to the right in the figure this would indicate that a certain amount of new capacity would be built. Once built, the relevant cost of this latest vintage equipment is not its unit total cost but its unit operating cost and it would be inserted in its appropriate place in the rising section of the supply curve. Thus at any instant, if current output is Q_0, the relevant supply curve for $Q < Q_0$ is the 'short-run' section to the left of Q_0 based on existing vintages of equipment. And for $Q > Q_0$ the relevant supply curve is the 'long-run' horizontal curve.

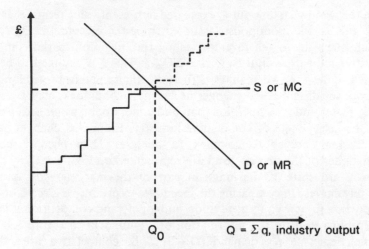

Figure 3.17: Simultaneous long- and short-run equilibrium

The picture is similar for the monopolist, except that the curve labelled D in Figure 3.17 is re-labelled MR (marginal revenue) and the curve labelled S is relabelled MC (marginal cost). In fact, with two extra assumptions the competitive supply curve is, except for a simple rescaling, identical with the MC curve. First, assume that the demand curve for the product is linear; then for any given level of long-run costs monopoly output is one half the corresponding competitive output. With the further assumption that the histories of technical change have been identical, which would be appropriate for exogenous technical change, the capacity of any previous vintage installed under monopoly will be one half that installed under competition. The diagram may then be used for both the monopoly and the competitive cases, with appropriate relabelling, observing that the scale of the abscissa is doubled for the former case.

It is now possible to derive the rewards to be gained from engaging in inventive activity. The invention may involve either an 'embodied' or a 'disembodied' technical change. The distinction is that an embodied change in technology refers to a lowering of unit total costs of new equipment only, whereas a pure disembodied change affects all vintages equally. Treating this latter case first, it would be represented by a uniform downward shift of the S (and MC) curve of Figure 3.17. If unit operating costs for all vintages are reduced by an amount ξ, expressed in present value terms, then the gain to the monopolist (assumed to be the inventor) is $Q_m.\xi$, while the gain to the inventor selling the invention to the competitive industry would be $Q_c.\xi = 2.Q_m.\xi$. This ξ is assumed to be equal under the two regimes if there is infinite patent protection. Hence for disembodied changes under these conditions, incentives are greater under competition than under monopoly, again because of the larger scale of competitive industry. Embodied changes in technology are more complex to analyse. They occupy the remainder of this section, and are represented on Figure 3.17 by a downward shift of the long-run part of the marginal cost and supply curves. In comparing the incentives to produce a given cost-reduction in new equipment under monopoly and competition it is helpful to transfer the information on Figure 3.17 to Figure 3.18 which depicts 'excess demand' $(D - S)$ for the competetive case and 'marginal profit' $(MR - MC)$ for the monopoly case.

The vertical axis of Figure 3.18 is the same as that of Figure 3.17 while the horizontal axis measures the difference between curve D (or curve MR) and curve S (or curve MC) on Figure 3.17. The curve $(D - S)$ connecting points ABCD ... is constructed from the horizontal difference between D and S on Figure 3.17. The curve $(MR - MC)$, connecting points AEFGH ... is obtained similarly from the horizontal difference between D and S (relabelled MR and

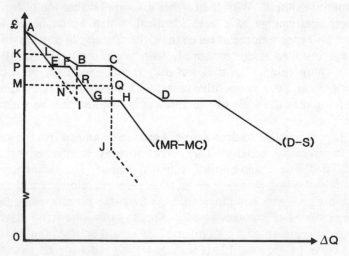

Figure 3.18: Excess demand and marginal profit functions

MC respectively), recalling that the scale of the horizontal axis on Figure 3.17 is doubled for MR and MC.

The curves (D – S) and (MR – MC) give on the horizontal axis the outputs of new equipment installed under competition and monopoly respectively corresponding to the marginal costs, which equals unit total costs, of such equipment on the vertical axis. Before the invention such marginal costs are given by OA. With a linear demand curve for the product, the curve (MR – MC) bisects the horizontal distance between the ordinate axis and the (D – S) curve.

Consider first the monopoly case. With a cost-reducing invention embodied in new equipment, the new level of long-run unit costs is given by a distance on the vertical axis such as OK or OM on the diagram. The extra profit to the monopolist as a result of the invention is given by the area ALK or area AEFRM respectively and the reduction in unit costs is AK or AM, respectively. In other words, as the new level of unit total costs is traced down the vertical axis from the initial point A, the area between the horizontal at that new cost level and the (MR – MC) curve gives the monopolist's revenue from the invention.

The curve $R^m(\xi)$ shown in Figure 3.19 gives the monopolist inventor's revenue from an invention that reduces unit operating costs of new equipment by ξ as a function of ξ.

The 'kink' in the curve corresponds to $\xi_1 = AP$ on Figure 3.18. A kink occurs whenever a vintage of installed equipment is

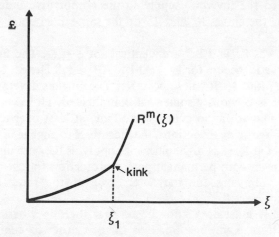

Figure 3.19: Monopolist's revenue function

rendered obsolete by the cost reduction. The slope of the curve is equal to the value of ΔQ on Figure 3.18 corresponding to the cost reduction. It measures the demand price of the invention and increases as ξ increases.

Turning now to the competitive case the curve $(D - S)$ of Figure 3.18 gives the effective demand for output produced by the latest vintage of equipment when the unit total costs of that vintage are given on the vertical axis. Treating $(D - S)$ as a demand curve in the fashion of section 3.2, the corresponding 'marginal revenue curve' is AIBCJ ..., which exhibits discontinuities between points I and B and between points C and J.

For any given post-invention unit total cost of production, the profit-maximising inventor will set a royalty rate such that he gains what would be monopoly profit with such a demand curve. Inspection of Figure 3.18 reveals an unusual problem in this setting in that at some values of 'marginal revenue' there may be two (or more) values of Q.

As ξ increases from zero the equilibrium point $(MC = MR)$ is traced out by the intersection of the new level of unit costs and curve section AI up to $\xi = AM$, say, at which point the area ANM is equal to the area PCQM (and equals the revenue to the inventor). For $\xi > AM$ the equilibrium point moves down curve section CJ and revenue to the inventor is given by the rectangular area bounded horizontally by the ordinate and CJ and bounded vertically by the new level of unit costs and PC. Labelling by R^c the revenue function to the inventor supplying the competitive industry and combining the curves R^c and R^m on the same diagram gives Figure 3.20.

The curves R^c and R^m are identical for $\xi \leqq \xi_1$. The first 'kink' in the curve R_c occurs for $\xi_2 = AM > AP = \xi_1$. Hence, for ξ lying between ξ_1 and ξ_2, R^m lies above R^c. The further development of the curves is obviously somewhat complicated. However, enough has been said so far to demonstrate that, at least over a range of small cost-reducing inventions, the incentives to engage in inventive activity are at least as high under monopoly as they are under competition, even with permanent patent protection for the inventor, and even allowing for the larger size of the industry under competition.

Finally, it must be acknowledged that there is a basic internal contradiction in this section. It is this: in order to analyse competitive and monopolistic product markets on the same footing it

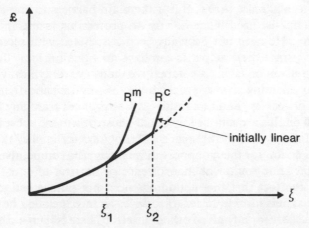

Figure 3.20: Competitive and monopolistic revenue functions

was assumed that both had the same history of purely exogenous technical progress; but if incentives to invent differ, as the analysis suggests, that fact is only interesting if there is a supply response of inventions by inventors – in other words, if technical progress has some endogenous component. This is inconsistent with the initial assumption of exogenous technical progress. The conclusion of this section is that with endogenous technical progress induced by incentives to invent there must have been different histories of technical progress in the two market structures unless it happened that the histories of output were identical – which could be the case if the monopoly were in a contestable market, implying that zero profit coincides with maximum achievable profit (see William Baumol *et al.*, 1982). And since, with embodied technical progress, the incentive to invent is a function of the past history of technical progress, which determines the current structure of costs, it must in general be the case that incentives will differ between the two market modes. But, intuition suggests, the difference may not be consistently in the same direction.

3.7 NON-PROFIT-MAXIMISING PRODUCTION MONOPOLY

Up to this point it has been assumed that where a production monopoly exists, and is presumably protected by entry barriers, its goal is to maximise profit. But if there are barriers to entry, then the monopolist need incur no cost to protect his property in the invention. He need not bear any costs associated with secrecy or patenting since these activities can have no value for him. By contrast, an inventor facing a competitive industry will typically incur costs in ensuring the appropriability of his invention through patents or secrecy, and as a result his net revenues from any invention will on this account be lower than those of a monopolist inventor. It follows that the existence of entry barriers implies that the profit function for the monopoly differs from the competitive case.

Another implication of the existence of barriers to entry is, of course, the fact that the monopolist need not pursue the goal of profit-maximisation in order to survive. In fact, he need not even minimise cost for any given output level.[19] There is then a plethora of alternatives to profit-maximisation as possible goals for the monopolist. To mention just a few that have been proposed: the monopolist may attempt to maximise the 'utility' of the management or even of the whole labour force (see Oliver Williamson, 1964); he may maximise the growth of the firm (see Robin Marris, 1964); he may maximise sales revenue subject to achieving a required minimum profit level (see William Baumol, 1962); or he may simply 'satisfice' (see Herbert Simon, 1955).

The utility-maximising models normally imply that the firm's management is prepared to sacrifice some profit to achieve other management goals such as employment or discretionary capital expenditures. The firm normally operates at an output rate greater than an equivalent profit-maximiser with the same cost and revenue functions. Since the sum the firm is prepared to pay for an invention (the incentive to invent) is at least as great as the increase in profit which, to a first approximation, is given by the reduction in costs times output, the utility-maximiser has greater incentive to invent than the profit maximiser.

In the remainder of this section the 'revenue maximisation subject to a profit constraint' hypothesis is analysed in some depth for its implications regarding the incentives to invent under monopoly. To begin with, an account of the basic revenue-maximising model is given. The implied incentive to invent is then derived.

Let $S(Q)$ denote the firm's total sales revenue as a function of the output rate Q. With $S(Q) = p(Q).Q$ and with $p'(Q) < 0$ reflecting the fact that the demand curve facing the firm slopes downwards, marginal revenue is positive and declines with increasing output: $S'(Q) > 0$; $S''(Q) < 0$.

Denoting by c_0 the firm's constant unit costs, the profit function is given by:

$$\pi = S(Q) - c_0.Q$$

and π_{max} is the maximum obtainable profit rate;
while π_{min} is the required minimum profit rate.
The standard results of the model are as follows:

(i) if $\pi_{min} > \pi_{max}$ then $Q = 0$;

(ii) if $\pi_{min} = \pi_{max}$ then $S'(Q) = c_0$;

(iii) if the profit constraint is binding, $\pi_{min} = \pi \leqq \pi_{max}$, then $S'(Q) \leqq c_0$; and

(iv) if the profit constraint is not binding, $\pi_{min} < \pi$, then $S'(Q) = 0$.

The firm's output only coincides with that of a profit-maximising monopolist if the required rate of profit equals the profit-miximiser's profits. If the required rate of profit is greater than maximum profits, then output is zero; if it is less than maximum profits, then output is greater then the profit-maximiser's output. The revenue-maximiser's output is less than competitive output unless both the required profit is zero and demand at price equal to the unit cost level is elastic, in which case the output rates are equal.

Turning now to the firm's incentives to engage in inventive activity, results (i)–(iv) above suggest that they will depend upon the required minimum rate of profit as well as the initial and post-inventive levels of unit costs.

The incentive to produce a given cost-reducing invention is defined as the value of the invention to the firm. It is the maxumum price that the firm is prepared to pay for the invention consistent with the condition that revenue afterwards is at least as great as pre-invention revenue. This suggests that if the profit constraint is not initially binding there is no value in a cost-reducing invention to the firm. There is no incentive to innovate since marginal revenue is already zero and further reductions in cost may increase profit but cannot serve to give greater revenue.

If the profit constraint is initially binding then a cost-reducing invention releases that constraint and may allow revenue to increase. The maximum price the firm is prepared to pay for the invention is the measure of the firm's incentive to invent, and in the limit the firm is prepared to forgo all the surplus profit released by the invention so long as revenue remains at least at the pre-invention level.

Denoting the pre- and post-invention output rates by Q_0 and Q_1 respectively, pre-invention profit is given by:

$$\pi_0 = \int_0^{Q_0} S'(Q).dQ - c_0.Q_0$$

and post-invention profit is:

$$\pi_1 = \int_0^{Q_1} S'(Q).dQ - c_1.Q_1$$

$$= \int_0^{Q_1} S'(Q).dQ - (c_0 - \xi).Q_1$$

Hence the profit released by an invention of size ξ is:

$$\Delta\pi = \int_{Q_0}^{Q_1} S'(Q).dQ + \xi.Q_1 - c_0.(Q_1 - Q_0)$$

For a given cost reduction of size ξ, $\Delta\pi$ is a function of Q_1 alone. The incentive to produce the given invention is then given by:

$$R^r = \max_{Q_1} \Delta\pi, \text{ such that } S(Q_1) \geqq S(Q_0)$$

where the superscript ^{r} reminds us that this is for a revenue-maximiser.

Setting up the Lagrangean for this non-linear programming problem:

$$L(Q_1; \lambda) = \int_{Q_0}^{Q_1} S'(Q).dQ + \xi.Q_1 - c_0.(Q_1 - Q_0)$$

$$+ \lambda.[S(Q_1) - S(Q_0)].$$

Differentiating with respect to Q_1 and λ, and applying the Kuhn–Tucker conditions gives:

$$\partial L/\partial Q_1 = (1 + \lambda).S'(Q_1) + \xi - c_0 \leqq 0$$
$$Q_1.[(1 + \lambda).S'(Q_1) + \xi - c_0] = 0$$
$$Q_1 \geqq 0 \qquad\qquad 3.9$$

and:

$$\partial L/\partial \lambda = S(Q_1) - S(Q_0) \geqq 0$$
$$\lambda.[S(Q_1) - S(Q_0)] = 0$$
$$\lambda \geqq 0 \qquad\qquad 3.10$$

Putting conditions 3.9 and 3.10 to use, consider first the initial situation corresponding to (ii) in the foregoing list of standard results, that the required minimum profit level is equal to the maximum obtainable profits under the inital conditions. The firm is in production 'by the skin of its teeth'.

Assume that the firm remains in production after the invention, i.e. that $Q_1 > 0$. The initial conditions give:

$$c_0 = S'(Q_0)$$

Conditions 3.9 amount to:

$$0 \leqq \lambda = (c_0 - S'(Q_1) - \xi)/S'(Q_1) \qquad\qquad 3.11$$

Now $\lambda = 0$ since assuming $\lambda > 0$ involves a contradiction.[20]

Hence, $\xi = c_0 - S'(Q_1)$, and the incentive to invent may be expressed:

$$R^r \equiv \Delta\pi = \int_{Q_0}^{Q_1} S'(Q).dQ + S'(Q_0).Q_0 - S'(Q_1).Q_1$$

which is given by the shaded area in Figure 3.21.

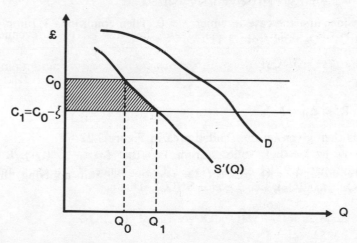

Figure 3.21: Incentive to invent with initially binding profit constraint.

As this is equivalent to the incentives for the monopoly inventor depicted in section 3.2 above, it can be seen that:

$$R^r(\xi) = R^m(\xi)$$

The incentive to invent under revenue-maximisation equals that under profit-maximisation if the required minimum profit rate equals the maximum attainable profit rate in the initial conditions.

If initial conditions (i) above apply, so that required profits are greater than the initially maximum attainable profits ($\pi_{min} > \pi_{max}$), then the initial output, Q_0, is zero and the cost reduction must be greater than the difference between unit required profit and unit maximum initial profit if it is to be of any value to the firm. With this condition met, the analysis above applies, and the incentive function may now be written:

$$R^r(\xi) = \begin{cases} 0 & \text{if } \xi.Q^m \leqslant \pi_{min} - \pi_{max} \\ R^m(\xi) - \pi_{min} & \text{if } \xi.Q^m \geqslant \pi_{min} - \pi_{max} \end{cases}$$

Now consider the 'standard case' of initial conditions (iii) above, in which the profit constraint is binding and marginal revenue is less than marginal cost but greater than zero, i.e.:

$$0 < S'(Q_0) < c_0$$

Again equation 3.11 holds:

$$0 \leqq \lambda = [\, c_0 - S'(Q_1) - \xi\,]/S'(Q_1) \tag{3.11}$$

Consider first the case in which $\lambda > 0$. Then conditions 3.9 imply that $Q_1 = Q_0$, and from 3.11:

$$\xi < c_0 - S'(Q_0)$$

and

$$R^r \equiv \Delta\pi = \xi.Q_0$$

R^r is then given by the shaded area in Figure 3.22.

Now let $\lambda = 0$. It follows from 3.11 that $\xi = c_0 - S'(Q_1)$. But conditions 3.9 give: $S(Q_1) \geqq S(Q_0)$, hence $Q_1 \geqq Q_0$, and $S'(Q_1) \leqq S'(Q_0)$, i.e. $\xi \geqq c_0 - S'(Q_0)$. Thus:

$$R^r \equiv \Delta\pi = \int_{Q_0}^{Q_1} S'(Q).dQ + c_0.Q_0 - S'(Q_1).Q_1,$$

which is shown as the shaded area in Figure 3.23

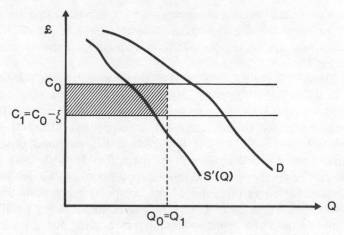

Figure 3.22: Initial and final profit binding constraints

Comparing Figures 3.22 and 3.23 with Figures 3.3 and 3.4, it can be seen that this 'standard' case of the revenue maximising model implies greater incentives to invent than under monopolistic profit-maximisation for all 'sizes' of invention, but that the incentives are less than in the competitive case, unless the profit constraint is zero, in which case they coincide.

In summary, the shape and position of the $R^r(\xi)$ curve depends upon the initial conditions (i)–(iv) outlined at the beginning of this section. The curves may be compared with each other and with that

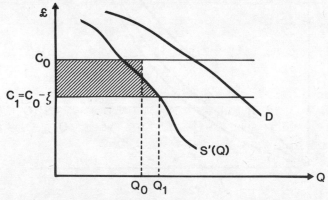

Figure 3.23: Invention removes initially binding constraint

for the profit-maximising monopoly, $R^m(\xi)$, as well as that for the inventor supplying a competitive industry, $R^c(\xi)$ in Figure 3.24, which labels the curves (i)–(iv), corresponding to the forementioned initial conditions.

In Figure 3.24 the curve labelled (iii) corresponds to the standard case in which the initial profit constraint is binding but the firm is in production. It is linear up to $\xi = \xi_1$, and thereafter convex. This may be compared to the curve for a competitive industry, R^c, which is linear up to ξ_2, the cost reduction corresponding to a drastic invention, and again convex thereafter. In both cases the slope of the linear section is equal to output: of the revenue-maximiser for curve (iii), and of the competitive invention-using industry for curve R^c. For $\xi > \xi_2$ the curves are vertically parallel.

Where initially the profit constraint was binding at the level of maximal profit, the firm behaves like a simple profit-maximising monopoly, and curve (ii) applies. It coincides with the curve R^m.

Curve (iv) is coincident with the abscissa, and corresponds to the case in which the profit constraint was initially non-binding.

If the profit constraint was initially binding and forced Q_0 to zero, then curve (i) applies: it is horizontally parallel to curve (ii), and as the profit constraint is relaxed (i) is shifted horizontally leftwards until it coincides with (ii). Further relaxation of the profit

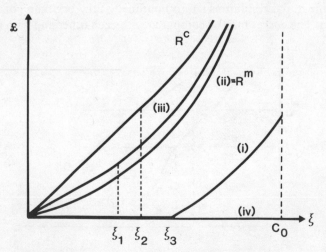

Figure 3.24: Incentives to invent compared

constraint transforms the curve to one such as (iii), with an initially linear section which grows in length and is rotated anti-clockwise as the profit constraint is further reduced until, in the limit as the required profits go to zero, it coincides with curve R^c.

It is clear from Figure 3.24 that generalisation about the incentives to invent under revenue-maximisation are difficult to make since it all depends upon the initial conditions. However, it may be noted that the 'standard case' in which initial output is positive and the profit constraint is binding, gives an intermediate incentives function between the profit-maximising monopoly and competitive cases.

3.8 CONCLUSION

This chapter has shown that the derived demand for an invention depends on the characteristics of the invention-using industry. Of prime importance is the size of the industry, but other factors such as market organisation and factor prices are also important determinants of an invention's social and private value. Additionally, the degree of appropriability of the invention, which was examined in terms of the duration of patent protection, and the short- or long-run nature of the invention-using industry's cost curve will be relevant considerations. This chapter has considered how the derived demand for a given invention depends upon these factors, but there may also be an influence from the conditions in the invention-using industry to the nature of the invention itself. The next chapter explores an important example of such endogeneity of invention characteristics, namely the influence of factor prices on the factor-saving bias of cost-reducing inventions.

NOTES

1. William Nordhaus (1969, Chapter 1) notes that although firms report only 13 per cent of research effort going into process inventions, the true figure is nearer 80 per cent when it is taken into account that much of the rest of industrial non-defence R & D is carried out in capital goods producing industries.
2. This is currently an active and important field of development in industrial economics. Recent advances include work by Avner Shaked

and John Sutton (1983; 1985), who demonstrate that a monopolistically competitive firm's optimal invention strategy for new products
depends crucially on the distinction between 'horizontal' and 'vertical'
(ie. quality) product differentiation, and for the latter on the way in
which the fixed cost of research can substitute for higher variable cost
in product improvement.

3. What prices should enter into the summation depends on whether the
   seller of the invention can price discriminate or not. If he cannot then
   not all potential buyers actually pay their reservation price. On the
   other hand, if price discrimination is possible then all buyers may pay
   their various reservation prices, but these will be interdependent: how
   much one buyer is prepared to pay for an invention depends on
   whether a rival might also acquire it.

4. Most of this literature has been concerned with the question whether
   competition or monopoly in the product market is the more
   favourable for invention. Underlying this is interest in a matter of
   practical policy: namely whether it provides a case for strengthening
   or attenuating policies designed to promote competition, which are
   founded on the static theory of resource allocation without reference
   to changing technology. Thus Arrow's (1962) assertion that a competitive product market product market provides greater incentives to
   invent than does monopoly appeared to strengthen the case for 'trust
   busting'. But Harold Demsetz (1969) pointed out that this followed
   from the normal output-reducing effect of monopoly, and that normalising the demand curves so that the pre-invention output would be
   the same under competition and monopoly has the effect of reversing
   Arrow's conclusion. Subsequent refinements of this normative
   analysis were presented by Morton Kamien and Nancy Schwartz
   (1970), Basil Yamey (1970) and S. H. Hu (1973).

5. Some explanation is really required for this long-run monopoly position, because it is not a natural consequence of the cost assumptions.
   A natural monopoly would be implied by a downward-sloping, long-
   run average total cost curve. An explanation could be that there is
   an initial fixed cost for being in production at all. This amount ought
   then to be deducted from the hatched areas in the diagram to show the
   true monopoly profit.

6. This distinction was first made by Arrow (1962), and has subsequently
   on occasion been emphasised by other writers employing this analysis,
   for example by Kamien and Schwartz (1982, Chapter 2) who point out
   that it is implicit in their unified treatment of monopoly and competition in a Cournot oligopoly framework.

7. Denote the demand curve by $q = q(p)$ with $q'(p) < 0$, where the prime
   signifies differentiation. The final price is the post-invention level of
   unit costs plus the per unit royalty rate: $p = c_1 + r$, and since price
   cannot exceed the initial level of unit costs, $p \leq c_0$, so $r \leq c_0 - c_1$. The
   royalty rate is determined by maximising total royalty payments,
   $r.q(p)$, with respect to r subject to this constraint. This is a problem
   in nonlinear programming, and the first order Kuhn–Tucker condi-

tions (see, for example, George Hadley, 1964) can be expressed:

$$r.(q + r.q' - \lambda) = 0 \text{ and } r \gtreqqless 0$$
$$\lambda.(c_1 - c_0 + r) = 0 \text{ and } \lambda \gtreqqless 0$$

Now if the constraint is binding (i.e. minor invention) then $\lambda > 0$ so $r = c_0 - c_1$ which implies $p = c_0$. But if the constraint is non-binding then $\lambda = 0$ which, with a positive r implies $r = -(q/q')$. At the dividing line both conditions hold, therefore $c_0 - c_1 = -(q/q')$ and:

$$(c_0 - c_1)/c_0 = -q/p.q' = 1/\varepsilon$$

where ε is the elasticity of demand.

8. See note 5 above.
9. This is Arrow's (1962) assumption, and it is maintained by Kamien and Schwartz (1982) in their survey of the literature.
10. Basil Yamey (1970) pointed out that the lump-sum form of payment is implicit in the analyses of the basic model presented by both Arrow (1962) and Demsetz (1969), since both assume that the monopolist's marginal cost curve is unaffected by payment for the use of the invention. Arrow had explicitly stated that the monopolist was himself the inventor, presumably to avoid the bilateral bargaining situation inherent in extramural inventions. Demsetz's dissection of Arrow's argument begins with an explicit assumption of extramural invention with the inventor denied the ability to discriminate in royalty rates – with the rate entering marginal costs just as it does for the competitive industry. It is only when Demsetz attempts to examine the case 'where rivalry between inventors, or where regulation fails to equalise the royalties' that the implicit lump-sum nature of the royalties slips in.
11. See Eugene Silberberg (1978, Chapter 7) for a good account.
12. Had the same procedure been carried out in terms of profit maximisation instead of cost-minimisation, then the implied 'unconditional' factor demands would be functions of the price of final output along with all factor prices. These are styled 'unconditional' since the level of output is endogenous. But unconditional factor demands cannot be defined for an important class of technologies; namely those for which the production function $f(x)$ is homogeneous of degree one. Therefore, since cost-minimisation is necessary, but not sufficient, for profit-maximisation, the conditional demand functions are implied by, but do not themselves imply, the corresponding unconditional functions.
13. See Hal Varian (1978, Chapter 1) for an exposition of the geometry of duality for cost and production functions.
14. Which of these is correct as a basis for comparison depends on the purpose. When normative issues are stressed, Demsetz (1969) points out that: 'monopoly models generally deduce that a monopolist will use less of all inputs, including an invention, because he produces less output; the demonstration of any *special* effect of monopoly on the incentive to invention requires that adjustments be made for this normally restrictive monopoly behaviour', and he accordingly prefers

the level of output to be given rather than the demand curve in this comparison.

15. This method of standardisation is due to Demsetz (1969). Further refinement of the methodology, in which allowance is also made for the elasticity of demand, is presented in Kamien and Schwartz (1970).

16. In fact the monopoly still has the edge over the inventor facing a competitive industry because his output increases. This will normally be a second-order effect, however, except where the elasticity of demand (or marginal revenue) is large, and the cost reduction substantial

17. If T is such that $D^m > D^c$ for all values of ξ, it must be true that:

$$D^c_\xi(\rho, L, T, 0) < D^m_\xi(\rho, L, 0)$$

or,

$$(1 - e^{-\rho T}).D^c_\xi(\rho, \infty, \infty, 0) < (1 - e^{-\rho L}).D^m_\xi(\rho, \infty, 0)$$

and since

$$\nu = Q^c/Q^m = D^c_\xi(\rho, \infty, \infty, 0)/D^m_\xi(\rho, \infty, 0)$$

it follows that:

$$(1 - e^{-\rho T}).\nu < (1 - e^{-\rho L}).$$

18. This qualification is designed to eliminate the case in which the only relevant cost is that of installed equipment.

19. This is implied by Harvey Leibenstein's (1966) concept of 'X-inefficiency', for example.

20. If $\lambda > 0$ then, from conditions 3.10:

$$S(Q_1) = S(Q_0)$$

which implies: $Q_1 = Q_0$

and hence, from 3.11:

$$\lambda = -\xi/S'(Q_0) < 0$$

which contradicts the initial assumption that $\lambda > 0$.

4 The Bias of Invention

The responsiveness of invention characteristics to factor prices is called the induced bias of invention. This chapter begins with a sketch of the recent history of ideas in this area. The relation of factor-saving bias to the cost function presentation introduced in section 3.4 above is outlined in section 4.2. This is followed by a brief summary of recent empirical evidence on the question of biased technical change. Section 4.4 introduces the idea of a trade-off in the invention production function between rates of factor augmentation. The trade-off is called an 'invention possibility frontier'. Section 4.5 addresses the question: what is the long-run tendency in the bias of invention? The final section recognises that the rate of technical change itself affects the demand for different types of invention when new technology is embodied in capital equipment.

4.1 INTRODUCTION

A number of ideas of seminal importance were introduced by Sir John Hicks in Chapter 6 of his *Theory of Wages* (1932). They include: the elasticity of substitution between factors of production; the classification of inventions by factor saving biases; the distinction between 'induced' and 'autonomous' invention; and the idea of a price-induced bias in the outcome of inventive activity. The seeds planted by Hicks still provide a flourishing research agenda today, half a century on.

The focus of Hicks's thinking is implied in the title of his Chapter 6: 'Distribution and Economic Progress'. The analysis is directed at the determinants of the relative shares of capital and labour in total income. The reasoning by which Hicks attempts to unravel this question is interesting. He argues that as that ratio of capital to labour is on a secular increase,[1] variations in relative shares over time are associated with changes in the overall elasticity of substitution: it being greater than one when capital's share is increasing, and less than one when labour's share is increasing. And a high elasticity of substitution is associated with a high level

91

of invention, especially induced invention, which will be labour-saving.

Hicks could consider the elasticity of substitution to be influenced by invention because, for him, the relation between outputs and inputs was not constrained to a given level of technology in the way that modern production functions are formulated.[2] In fact the production relation for Hicks is equivalent to what Vernon Ruttan and Yujiro Hayami (1971) refer to as a 'meta production function', which is alternatively dubbed the 'innovation possibility curve' (Syed Ahmad, 1966), a specialisation of which is Charles Kennedy's (1964) 'innovation possibility frontier', which in turn gives rise to the 'isotech' of William Nordhaus (1973).

A production function expresses the relation between outputs and inputs for a given technology – for a given state of the productive arts, that is – whereas an innovation possibility curve gives the outer envelope of all possible production functions attainable with a given amount of invention. If no invention takes place, the production function isoquant and the innovation possibility curve coincide. The difference between them is therefore due to the supply of inventions. But here, unlike in Chapter 2, the supply of inventions means not only its overall rate (i.e. amount of cost reduction) but also its directional bias.

It was argued in section 3.4 above that the demand price of an invention, and hence the inducement to invent, must in general be a function of relative factor prices. This has given rise to a theory of induced bias in technical progress. But while theories of induced bias have been popular as vehicles to analyse the direction of technical progress, they are not without their critics. In particular, W. E. G. Salter (1960) objected to the idea that dearer labour stimulates the search for new knowledge aimed specifically at saving labour because:

the entrepreneur is interested in reducing costs in total, not particular costs such as labour costs or capital costs. When labour costs rise any advance that reduces total cost is welcome, and whether this is achieved by saving labour or capital is irrelevant. There is no reason to assume that attention should be concentrated on labour-saving techniques, unless, because of some inherent characteristic of technology, labour-saving knowledge is easier to acquire than capital-saving knowledge. (Chapter 3, pp. 43–4)

But Salter must be assuming here that invention characteristics are exogenous, that researchers are either unable to influence the type of inventions they produce or unable to distinguish the factor saving characteristics of different potential inventions. If he is correct in these assumptions then indeed there can be no induced bias to inventions, properly speaking, and any actual bias would be the result of purely technological factors uninfluenced by the profit-seeking of inventors or innovators. However, the viewpoint of the present chapter is that these are matters worth examining rather than rejecting out of hand.

4.2 THE DERIVED DEMAND FOR INVENTION CHARACTERISTICS

It was argued in Chapter 3 that the demand price of a cost-reducing invention is a function of the output of the industry to which the invention will be applied and of the prices of the factors of production employed by that industry. Basically this is because cost is itself a function of these arguments. Thus:

$$V(w, q) = \max\{C^0(w, q) - C^1(w, q), 0\} \qquad 4.1$$

where $V(\ )$ represents the demand price of the invention, $C^0(\ )$ is the cost function for the existing technology, and $C^1(\ )$ is the cost function for the technology given by the invention. If the production function is homogeneous of degree one, this can be specialised to:

$$V(w, q) = q.\max\{c^0(w) - c^1(w), 0\} \qquad 4.2$$

where $c(w)$ is the unit cost function. The development in Chapter 3 focused on the functional dependence of the demand price of invention on output, q. Discussion of the role of the vector of factor prices w was deferred to this chapter.

There is, in principle, an infinite number of ways in which a given reduction in costs, or unit costs, could occur. An example was given in Chapter 3 of the invention of a new technology of a type totally different from the old one, meaning that the new isoquant map was not a simple transformation of the old one. In the face of this plethora of possibilities it has been found useful to devise a categorisation of technical change in terms of the relative factor intensities and factor price ratios implicit in the differing

technologies. Accordingly, a 'Hicks-neutral' technical change is
one in which constant factor intensities in production imply and are
implied by constant factor price ratios. And a technical change is
biased if factor intensities change while factor price ratios stay constant, or if factor price ratios change while factor intensities remain
constant. Thus if a new technology implies that when the ratio of
wage rate to rental price of capital stays constant the ratio of
capital to labour falls, such a change exhibits capital saving bias in
the Hicks sense. Hicks-neutrality implies a radial inward shift of
the isoquant map, preserving the slopes of the shifting isoquants
along the rays. Equivalently, it implies a radial outward shift of the
factor price frontier, again preserving the slopes along the rays.

It is important to note that the Hicksian classification of
technical change is strictly local. It compares a particular point on
one isoquant with the corresponding point on the new one − corresponding either in the sense of having same factor intensities, or
in the sense of having the same slope. Thus in Figure 4.1(a) the
movement from point A on the original isoquant to point B on the
new one represents a Hicks-neutral change, but the movement from
the old point C to the new point D represents a capital saving bias
in the Hicks sense. The same movements are depicted on the corresponding price frontier in Figure 4.1(b). But while the Hicksian
classification scheme only compares points on the old and new iso-

(a) Isoquant   **(b) Factor price frontier**

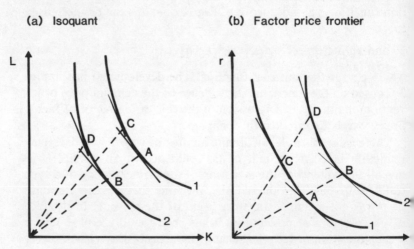

Figure 4.1: Hicks-neutral and capital-saving inventions

quants, or price frontiers, it is possible for the whole isoquant, or price frontier, to move globally in like manner so that the same Hicksian classification applies unambiguously for all factor prices or factor intensities. For analytical purposes it is often convenient to make the assumption that an invention represents such a globally Hicks-neutral or Hicks-biased change in technology. If, moreover, it is assumed that the form of the unit cost function remains unaltered, (though, of course, the value of the function must be changed for given arguments) then the invention is tantamount to an increase in the amount of efficiency of some factor or factors of production. This is known as 'factor-augmenting technical change'. It is assumed in the remainder of this chapter.

Factor-augmenting technical progress is usually represented in terms of the production function by the introduction of n terms, b_1, representing the efficiencies of the n factors of production, x_i:

$$q = f(b_1x_1, b_2x_2, \ldots, b_nx_n). \qquad 4.3$$

Thus where x_i measures the amount of the ith factor in natural units, b_ix_i measures it in 'efficiency units'. It is clear therefore that if the natural factor price is w_i then the price per efficiency unit is simply w_i/b_i. A factor augmenting technical change is now simply a new set of efficiencies for the factors in the production function. If the function f( ) displays constant returns to scale, then the unit cost function may be written:

$$C/q = c(w_1/b_1, w_2/b_2, \ldots, w_n/b_n) \qquad 4.4$$

and an increase in the efficiency of the ith factor, b_i, is equivalent to a fall in that factor's price w_i for a constant level of its efficiency.

The cost function 4.4 implies an optimal demand function for the ith factor which, on application of Shephard's lemma, is:

$$x_i = \partial C/\partial w_i = q.c_i/b_i$$

where $c_i \equiv \partial c/\partial(w_i/b_i)$.

If the demand for a factor is expressed in terms of efficiency units of that factor, it is:

$$b_ix_i = q.c_i \qquad 4.5$$

Now if factor prices are held fixed, but an invention implies small changes in the factor efficiencies, db_i, the consequent change in total costs dC gives the demand price for a differential invention:

$$v = - dC = - \sum_i (\partial C/\partial b_i).db_i = q.\sum_i (c_i w_i/b_i{}^2).db_i$$

$$= \sum_i x_i w_i \hat{b}_i \qquad 4.6$$

where $\hat{b}_i \equiv db_i/b_i$ is the rate of change of the ith factor's efficiency. (Note from 4.6 that the demand price of a differential factor augmenting invention is a weighted sum of the proportional changes in factor efficiencies, where the weights are expenditures on the relevant factors.)

For an invention that implies finite increments in factor efficiencies, instead of the infinitessimal differentials assumed to derive equation 4.6, the change in cost representing the derived demand for the invention can be shown pictorially as in Figure 4.2. The figure assumes that only the efficiency of the ith input is increased by the invention. This has the effect of reducing that input's effective price, and, with all else given, thereby reducing unit costs from point A to point B. The shaded areas are equal, and measure the

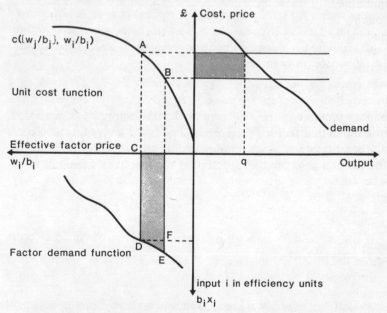

Figure 4.2: Derived demand for a factor-augmenting invention

derived demand for the invention. To see that the lower shaded area corresponds to the formula 4.6 for small changes in b_i, note that with only $db_i > 0$ and all other $db_j = 0$, equation 4.6 implies that $v = (q.c_i w_i/b_i^2).db_i = (b_i x_i w_i/b_i^2).db_i = (b_i x_i).(w_i/b_i^2)db_i$ where the first term in the final expression gives the vertical distance CD on Figure 4.2, and the second term simply $d(w_i/b_i)$, the distance DF. Therefore, the differential approximation 4.6 differs from the correct value, which allows for changes in the volume input of factor i, by the area DEF. When more than one factor efficiency changes, the unit cost curve itself shifts and the figure cannot be so easily interpreted.

4.3 EMPIRICAL EVIDENCE ON BIASED TECHNICAL CHANGE

Taking the grand view – that of the national economy over a long period of time – Sir John Hicks, and somewhat later Sir Roy Harrod, were able to form clear opinions about the factor-saving biasedness of progress. Both agreed that factor shares had been roughly constant over long periods and that the volume of capital had increased faster than labour. For Hicks this implied, under his own classification, that progress had been biased towards labour-saving innovations. But for Harrod the macroeconomic evidence was consistent with his own version of neutrality, which is defined in terms of a constant marginal product of capital and which can be interpreted as pure labour augmentation. Nowadays it is recognised that the matter involves a distinction between movements along a production function and shifts of the function itself. The former can be described in terms of the elasticity of substitution, whereas only the latter concern the characteristics of technical progress. However, while this distinction is important it cannot always be put into effect. Thus if the maintained hypothesis specifies a Cobb–Douglas production function, which has a unit elasticity of substitution, all movements of the function appear neutral under either definition of neutrality.

In recent years there have been several econometric attempts at simultaneous estimation of the form of the production or cost function and the factor saving bias of technical change. Most of these utilise the duality relation between production and cost functions, and in particular Shephard's lemma which states that the conditional demand function for a factor of production is the partial derivative of the cost function with respect to that factor's price. Also, most studies assume that the cost function has a flexible form; that is, a form consistent with the basic propositions of cost theory but which does not impose arbitrary restrictions on the parameters of the function. The econometric estimation of the parameters of the cost function is carried out by estimating either factor demand functions or factor share functions. In either case the principal explanatory variables are factor prices.

The biasedness of technical change is normally modelled by an assumption that it takes the factor-augmenting form. In terms of cost functions, this means that effective factor prices are declining over time, where the rate of decline equals the rate of factor augmentation. It is typically assumed that these proportional rates of decline are constant for any given factor, but differ between factors. The estimates of these rates of factor augmentation or price decline are used in conjunction with the estimates of substitution elasticities between factors to establish the bias of technical change on Hicks's classification of bias.

The typical starting-point for empirical analysis is the specification of a cost function:

$$C = C(q, \{w_i\}, t)$$

in which the term in braces, $\{ \}$, indicates the whole set of factor prices and t represents time − a proxy for technical change. The factor-augmenting specification would be:

$$C = C(q, \{w_i/b_i(t)\}).$$

A frequently used specification of the function is the translog form:

$$\ln C = \ln q + \beta_0 + \sum \beta_i.\ln w_i + \frac{1}{2}\left(\sum_i \sum_j \beta_{ij}.\ln w_i.\ln w_j\right)$$
$$+ \beta_t.t + \gamma_t.(t)^2 + \sum_i \gamma_i.\ln w_i.t$$

In this specification the β_i and β_{ij} coefficients summarise substitution possibilities while the γ_i coefficients represent the (constant, exogenous) bias of technical change towards the ith factor (factor i saving). The estimation of the substitution and bias parameters is carried out via factor demand or factor share equations. These are derived from the cost function by Shephard's lemma:

$$x_i = \partial C / \partial w_i$$

and

$$s_i = w_i x_i / \sum w_i x_i = \partial \ln C / \partial \ln w_i$$

where x_i denotes factor demand, and s_i represents the share of the ith factor in total costs. The attraction of these functions is that certain flexible forms of the cost function give rise to linear systems of factor demand or factor share equations, from which the original parameters are easily recovered. Thus the translog cost function is estimated by factor share equations:

$$s_i = \beta_i + \sum_j \beta_{ij} . \ln w_i + \gamma_i . t.$$

Another flexible form is the 'generalised Leontief' cost function:

$$C = q \sum_i \sum_j \beta_{ij} (w_i w_j) + q^2 \sum_i \beta_i w_i + qt \sum_i \gamma_i w_i$$

which yields the estimating equations:

$$x_i / q = \sum \beta_{ij} \sqrt{(p_j / p_i)} + \beta_i q + \gamma_i t$$

Most empirical studies along these lines have used United States data, though there have been studies using Swedish and Indian data. Also, the studies concentrate on post-war manufacturing industry, either in the aggregate or as a selection of its component industries. Their findings regarding factor-saving bias, are now summarised below.

Table 4.1 catalogues eight recent studies and the datasets they used, while Table 4.2 reports on their methods. The findings of these studies are now summarised briefly.

David and van de Klundert estimate a long-run elasticity of substitution between labour and capital of 0.32 and also estimate that the average rate of labour augmentation exceeded the rate of capital augmentation by 0.7 percentage points per annum over 62 years. They therefore conclude that technical change has been labour-saving, which is consistent with the induced innovation thesis since the price of labour has risen much faster than that of capital. This is corroborated by Binswanger, who finds the evidence strongly supportive of the induced innovation hypothesis. Thus he finds that the relative price of labour rose 60 per cent between 1912

Table 4.1

Study	Authors	Dataset
1	David and van de Klundert (1965)	US private sector, 1900–62
2	Binswanger (1974; 1978)	US agriculture, 1912–62
3	Bergstrom and Melander (1979)	9 Swedish manufacturing industries, 1950–73
4	Berndt and Khaled (1979)	US manufacturing, 1947–71
5	Wills (1979)	US primary metals industry 1947–71
6	Moroney and Trapani (1981)	12 US manufacturing industries, 1959–74
7	Jorgenson and Fraumeni (1981)	36 US manufacturing industries, 1958–74
8	Lynk (1982)	Indian manufacturing industries 1952–71

and 1968 while the labour saving bias of technical change was about 30 per cent over the same period. Similarly, the relative fertiliser price fell to one-eighth its original level while the fertiliser-using bias more than trebled (i.e. the share of fertiliser in total costs was over three times what would have been expected on the basis of substitution response to the price change alone). But the bias towards machinery (its share doubled over what substitution alone

Table 4.2

Study	Production function	Duality	Special features
1	Constant elasticity of substitution (CES)	No	Partial adjustment
2	Translog	Yes	Also cross-section data
3	CES	No	Partial adjustment
4	Generalised Box–Cox	Yes	Economies of scale allowed
5	Translog	Yes	Adds tfp* equation
6	Translog	Yes	Adds unit cost equation
7	Translog	Yes	Concavity of unit cost function ensured
8	Generalised Leontief	Yes	Homogeneous product industries

*tfp = total factor productivity

would predict) in the face of a 50 per cent rise in its relative price is, according to Binswanger, explained by the fact that innovation possibilities themselves are not neutral.[3]

The Swedish evidence of Bergstrom and Melander is also consistent with the induced innovation hypothesis as they find technical change was labour-saving in eight of their nine industries, and real wage costs considerably outpaced the price of capital services in Sweden over this period. This is supported by the results for the United States reported by Wills and by Moroney and Trapani. Both studies find the pattern of factor-saving to be correlated with relative price rises. The other two US-based studies, those of Berndt and Khaled and of Jorgenson and Fraumeni, are however only lukewarm towards the induced innovation hypothesis. Berndt and Khaled attribute more importance to economies of scale than to technical change, though they do find the bias of technical change to be correlated with relative price changes in the manner predicted by the induced innovation hypothesis. Jorgenson and Fraumeni contradict all other studies in their finding that technical progress has had a labour-using bias. Finally, the results reported by Lynk for selected Indian manufacturing industries finds a capital-using bias but neutrality as far as labour is concerned. Though Lynk does not present evidence on relative price movements, his findings are plausible in terms of the induced innovation hypothesis since the wage–rental ratio has almost certainly not moved so sharply upwards in India as in the developed western economies.

In summary, the evidence from these empirical studies is broadly consistent with the induced innovation hypothesis, though there is some room for doubt on the issue. The remainder of this chapter reverts to theory, taking it for granted that the bias of technical progress is to a certain extent determined endogenously by considerations of profit, as is implicit in the notion of induced bias.

4.4 THE INNOVATION POSSIBILITY FUNCTION

The innovation possibility frontier is a putative functional relation between the rates of change of the efficiency factors, b_i, of the factor-augmenting production function 4.3. It was introduced in the context of equilibrium macroeconomic growth theory by

Charles Kennedy (1974), when it appeared to offer a means by which a rather awkward conundrum in growth theory could be solved, or at least side-stepped, see Paul Samuelson (1966) or E. M. Drandakis and E. S. Phelps (1966), for example. There was a widely held belief at the time that long-run macroeconomic growth could be reasonably described by steady-state growth models in which, among other constancies, factor shares are constant. This particular constancy had even acquired a name: Bowley's law. It appeared to be one of the 'stylised facts' of economic growth. But the theory of growth implies that there is only one form of exogenous technical progress that can be admitted by such steady-state models, viz. 'Harrod-neutral' progress, which is the special case of factor-augmenting technical progress when all the b_i other than that attached to the labour argument are constant over time.[4] This, in essence, was the conundrum.

The history of classification of invention has concentrated on the effects of inventions on factor shares. Thus an invention has been described as 'neutral' if its application leaves factor shares unaffected. As it stands this constraint is insufficient completely to characterise the direction of technological change, but if the path along which factor shares are held constant is also specified, then a taxonomy of inventions is possible. Thus what is known as Hicksneutral technical progress keeps factor shares constant when either the capital–labour ratio (assuming these two factors only) is constant or, what amounts to the same thing, when the factor price ratio is held constant. And Harrod-neutral technical progress refers to a path along which the output–capital ratio, or equivalently the marginal product of capital services, is constant. The different nature of Hicks- and Harrod-neutrality is illustrated in Figure 4.3 in terms of unit isoquants and factor price frontiers. Another type of neutral technical progress is 'Solow-neutrality', which is similar to Harrod-neutrality, except that the roles of capital and labour are interchanged.

In the special case of a factor-augmenting production function, such as that given by equation 4.3, a straightforward account of the different principles by which an invention is classified as either labour-saving or capital-saving under the three types of neutrality mentioned above is possible. For two factors, capital K and labour L, the factor-augmenting production function takes the form:

$$Q = F(A(t)K, B(t)L)$$

(i) Isoquants

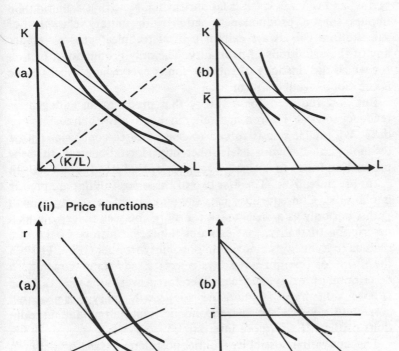

Figure 4.3: Hicks-neutrality (a) and Harrod-neutrality (b)

where A(t) and B(t) are the efficiencies of capital and labour respectively, for which a and b now denote rates of change. Defining the elasticity of substitution of the function F(.,.) as:

$$\sigma = F_K F_L / F F_{KL},$$

where the subscripts denote partial derivatives, the bias of technical progress, D, can be expressed

$$D = [(1 - \sigma)/\sigma] . \Psi(a,b) \qquad 4.8$$

where for the Hicks classification, $\Psi(a, b) = b - a$; for the Harrod classification, $\Psi(a, b) = b$; and for the Solow classification, $\Psi(a, b) = a$. Edwin Burmeister and Rodney Dobell (1970, Chapter 5) give details for the derivation of equation 4.8.

When D = 0 technical progress can be described as neutral under the relevant classification; when $D > 0$ technical progress is capital-saving; and when $D < 0$ it is labour-saving. Note that definition 4.8 implies that a production function with unitary elasticity of substitution will always exhibit neutral technical progress under any of the definitions of neutrality. The only production function for which the elasticity of substitution is everywhere equal to one is the Cobb–Douglas form.

But now the need for a theory that produces an endogenous tendency towards Harrod-neutrality in macro growth models is evident. Without it one is forced to one of three conclusions: that technological change is itself inherently Harrod-neutral; that the production function is Cobb–Douglas; or that steady-state growth paths are unrealistic. The first two of these possibilities are plainly implausible. Consequently, it is the attachment to steady growth paths, not only as a benchmark for more complex theory but as a description of reality, that explains the recent history of thought that led to Kennedy's innovation possibility frontier (IPF). Though the thrust of opinion nowadays rejects steady-state growth as a description of reality, and leans therefore towards the view that the IPF was something of a blind alley for growth theory, it is probably correct to say that the microeconomic implications and foundations of the IPF are more interesting.

The innovation possibility frontier postulates a trade-off between achievable rates of labour- and capital-augmentation, and may be written thus:

$$a = g(b) \text{ with } g'(b) > 0, \quad g''(b) < 0, \quad g(0) > 0.$$

This says that the attainable rate of capital augmentation (efficiency change of capital), a, is inversely related to the attainable rate of labour augmentation (efficiency change of labour), b, and to get more of the one implies increasing marginal sacrifices of the other. The function is sketched in Figure 4.4.

By postulating the existence of such a function, unchanged at each point in time, available to competitive firms in the economy whose objective is to maximise the instantaneous rate of cost reduction, it was possible to show that an economy with a constant savings ratio will asymptotically experience Harrod-neutral technical progress if the elasticity of substitution of labour for capital is less than one. With the elasticity of substitution greater than one, a knife-edge problem is encountered, namely that, except

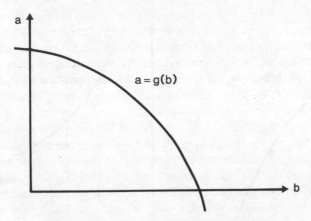

Figure 4.4: The invention possibility frontier

along one particular path, the share of capital either increases or decreases continuously, depending on the starting-point. But the case when the elasticity of substitution is less than one provided the greatest interest as far as the development of the theory of induced innovation is concerned because, without *assuming* Harrod-neutrality, this particular form of technical change is nonetheless experienced in the models – at least asymptotically.

One of the problems with Kennedy's IPF is that it was just postulated to exist, without much justification from microeconomics or any other source. It appears to firms exogenously and costlessly – no resources are required to achieve any of the factor augmentation mixes it depicts, i.e. it acts as a *deus ex machina* (Tobin, 1967).[5] Another problem is the myopic objective function assumed for the competitive firms, where survival would really require them to minimise the discounted value of all future costs. This is especially important if capital cannot be costlessly 'remoulded' like jelly, or if factor prices are expected to change over time. Finally, the stationarity of the function, independent of the time path firms or the economy experience, is rather limiting.

Why should the IPF have the concave shape sketched in Figure 4.4? The answer may become transparent if the IPF is transformed to what Nordhaus (1973) calls an 'isotech'. This is shown in Figure 4.5, the lefthand side of which shows an IPF with some details for particular points. The axes of the diagram on the right are those a

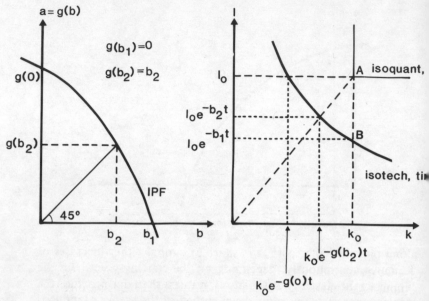

Figure 4.5: The isotech derived from the innovation possibility frontier

unit isoquant diagram, where $1 = L/Q$ measures unit labour requirements and $k = K/Q$ unit capital requirements. The current production point is at A, the corner, say of a Leontief fixed coefficients isoquant, drawn in as an L shape. The isotech is the locus of possible unit factor requirements attainable from invention at time t in the future, constructed by reference to the IPF on the left. For example, if technical change is purely labour-augmenting, $b = b_1$ and $a = g(b_1) = 0$, giving the next period's production point (at time t) at B, vertically below A on the isotech.

The isotech is, in a sense, an *ex-ante* isoquant. It tells what combination of inputs are possible at some future time as a result of the opportunities given by the innovation possibility frontier. Unit input requirements at time t are:

$$k(t) = k_0.\exp(-g(b)t) \text{ and } l(t) = l_0.\exp(-bt) \qquad 4.9$$

Differention of these expressions, with t fixed at unity, yields:

$$dk/dl = (k/l).g'(b) < 0$$

$$d^2k/dl^2 = (k/l).(g'^2 - g'') > 0,$$

which account for the negative slope and convexity of the isotech.

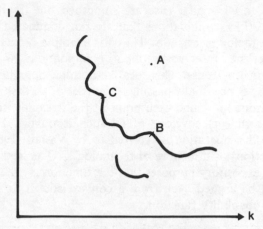

Figure 4.6: Irregular isotech

Obviously, as in the case of an isoquant, the cost-minimising unit input requirements at the future time to which the isotech relates are given by equality of the factor price ratio to the absolute slope of the isotech. The analogy with the traditional isoquant is complete. But why should the curve be smoothly convex? This is a problem posed by Nordhaus (1973). He expresses his disquiet by sketching what he terms an 'irregular' case, as shown in Figure 4.6.

It is easily demonstrated that a large number of 'bumps' and

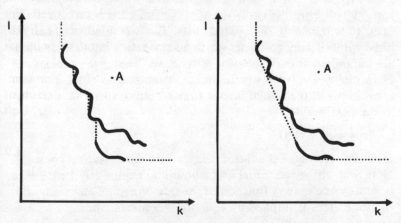

Figure 4.7: Convexification of an irregular isotech

'islands' in this irregular case are smoothed out if attention is restricted to (l,k) combinations that are not dominated for every conceivable factor price ratio. Thus for example points B and C would never be observed as input combinations as they are dominated in this sense. This local smoothing operation can be conducted by conceptually passing the corner of an L-shaped fixed coefficient isoquant round each point of the irregular isotech and forming the (linear) envelope of all such isoquants. Figure 4.7 illustrates. This operation is referred to by Varian (1980) as the 'convexification' of a mixture of technologies. The next step is to assume for expository purposes that the function is differentiable. It can then be argued back from a convex isotech to a concave innovation possibility function.

4.5 THE TENDENCY TO NEUTRAL TECHNICAL CHANGE

The objective now is to examine the implications of an exogenous, stable IPF for the labour/capital ratio and the rates of factor augmentation over time to establish whether a firm eventually moves towards a 'neutral' mode of technical change.

In line with the standard assumption of macro-induced innovation theory, the firm's objective is taken to be minimisation of instantaneous cost. Of course, this assumption is inappropriate if fixed capital is a sunk cost. In that case is would be necessary to formulate the problem in terms of maximising the net worth of the firm, or minimising discounted costs. But where services can be hired without sunk costs, net worth maximisation implies that costs are minimised at each moment of time, so the choice problem can be formulated in terms of minimising instantaneous costs. The firm is to choose b, the rate of labour augmentation, such that unit cost in the next period:

$$c = wl(t) + rk(t)$$

is minimised, where the factor prices w and r are assumed constant. Note that with given initial unit labour and capital requirements, c is, at any one time, a function of the rate of capital augmentation, b, alone. This is implied by equation 4.9 above, thus:

$$c = rk_0.\exp(-g(b)t) + wl_0.\exp(-bt) \qquad 4.10$$

Differentiating with respect to b and equating to zero, it can be seen that the optimal point is where the IPF has a slope equal to the factor share ratio:

$$g'(b) = -wl/rk \qquad \text{4.11}$$

This condition implies a minimum since the second order derivative is positive. From the first order condition 4.11, it follows that:

$$db/d(l/k) = -w/rg'' > 0 \qquad \text{4.12}$$

which implies that as labour intensity rises, so too does the optimal rate of labour augmentation. Noting that the rate of change of labour intensity can be written:

$$d(l/k)/dt = (l/k).(g(b) - b) \qquad \text{4.13}$$

and combining this with 4.12 yields:

$$db/dt = -wl.(g(b) - b)/rkg''(b)$$

Therefore,

$$db/dt \lesseqgtr 0 \text{ as } g(b) \lesseqgtr b \qquad \text{4.14}$$

Denote the difference between rates of factor augmentation by $H(b) \equiv g(b) - b$, then it follows that:

$$dH/dt \lesseqgtr 0 \text{ as } H \lesseqgtr 0 \qquad \text{4.15}$$

In words, propositions 4.14 and 4.15 state that if the rate of capital augmentation is greater (less) than the rate of labour augmentation then the labour-capital ratio is rising (falling) and the difference between the rates of capital and labour augmentation is falling.

Together they imply an equilibrium in which the rates of factor augmentation are equal and the labour capital ratio is unchanging through time.

Asymptotically we find, therefore, that the system converges towards a state of Hicks-neutral technical change. This may be depicted in a number of different ways, as Figures 4.8, 4.9 and 4.10 illustrate. The downward sloping curve in (l/k,H) space in Figure 4.8 corresponds to equation 4.13. The firm operates at each point in time on the curve and the arrows indicate the direction of movement of the variables through time. A starred variable denotes its equilibrium value.

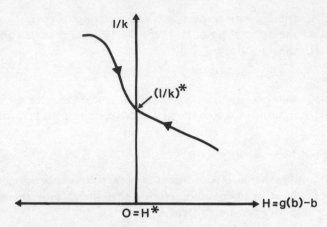

Figure 4.8: Equilibrium tendencies of factor intensity

An alternative representation, in terms of the innovation possibility function, is given in Figure 4.9. This illustrates the substance of equation 4.15, and the inference that in long-run equilibrium the rates of factor augmentation will be equal.

The picture is more complex if an attempt is made to depict it on the isotech diagram. The problem, of course, is that the shape and position of each period's isotech depends upon the (k,l) combination at the beginning of the period, which changes from period

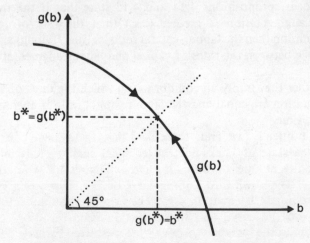

Figure 4.9: Equilibrium tendencies along the IPF

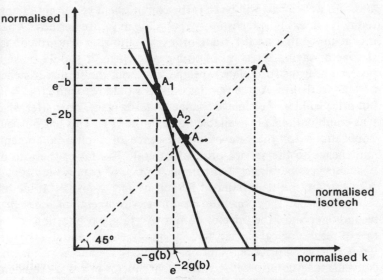

Figure 4.10: Tendency to equilibrium along the isotech

to period. One way round this problem is to make the isotech invariant from period to period by normalising the variables k and l such that they equal 1 at the beginning of each period. This implies that the slope of the isocost curves in the normalised variable is changing from period to period. In fact if the slope in the first period is equal to β then the slope at time t equals $\beta.\exp(-H(b)t)$. The term $\exp(-H(b)t)$ is easily depicted on the normalised diagram as the slope of the ray from the origin to the equilibrium point at the end of the first period. Taking $H(b) > 0$ for illustrative purposes, it can be seen that the next period's normalised isocost curve is less steep than that of the previous period. This is shown in Figure 4.10.

The equilibrium points move along the isotech from A_1 to A_2 and so on until, asymptotically, A_∞ is approached at which point technical change is equally factor-augmenting.

4.6 INDUCED PRICE EXPECTATIONS

Today's choice of factor input combinations may be affected by expectations about future movements in the prices of those inputs and of output. This will be so if the commitment to certain factors today is not easily and costlessly reversible in future periods. When it is assumed that an IPF trade-off exists, allowing firms to choose the factor-saving direction of technical change, it should be supposed that firms formulate expectations about the future existence of such tradeoffs. At the very least firms would recognise that output price will fall if technical change is taking place, no matter what bias combinations are available through the IPF. This has obvious implications for the rate of obsolescence of capital equipment, and thence to the service price of capital. The faster the rate of obsolescence the higher is the implicit price of capital services.[6]

In this section it is assumed that firms facing an IPF trade-off recognise that they operate in an environment of changing technology. Since the price of output will fall in proportion to the rate of technological change, and the quasi-rent of capital — the difference between price and unit operating costs — along with it, the convenient assumption of section 4.5 that factor prices are constant cannot be maintained. In other words, the factor price ratio relevant to a choice having implications for the future and the position of the IPF or isotech are not independent.

Assuming a given rate of price decline, it is in principle possible to deduce at what time current capital will become obsolete in terms of the various parameters. The problem then becomes one of choosing unit factor requirements (l,k), or equivalently the rates of factor augmentation, such that the present value of an investment is maximised subject to the innovation possibility trade-off function.

The model to be used is a variant of Salter's (1960) model, giving *ex-ante* substitutability between factors and *ex-post* fixity. Once capital has been committed, the unit capital and labour requirements in production with this equipment are fixed. A choice however is allowed the firm in decision-making before the capital is built, represented in the present analysis by an innovation possibility function or isotech.

It is assumed that the supply price of finance, that is, the rate of interest, and the asset prices of capital goods remain fixed, though the price of capital services may still vary as a result of changing expectations about the speed of technical progress. Initially these too are taken as given, so that the firm's choice problem, when it is confronted by an innovation possibility frontier, involves a given factor price ratio.

On the isotech diagram the intercept on the l axis of any given isocost curve represents unit labour cost measured in wage units if the current production point is located on that isocost curve. Assuming that the most efficient technique in production to date is represented by point A in Figure 4.11 and that the market factor price ratio is given by the slope of the line BAC then the distance OB is a measure of current unit total costs, and, following Salter and invoking the assumption of competition, it also measures the current price of output in (constant) wage units, p_0/w. The isotech represents the combination of new techniques available for the investment of the next period. Output will still carry on using technique A in the next period so long as there is an operating surplus with price greater than unit wage costs using that technique (i.e. $p \geqq wl_0$). But technique A will cease to be used when it becomes obsolete ($p < wl_0$).

However, in deciding on additions to the capital stock in the current period the firm may choose any point on the given isotech. Any point so chosen will have the implications for current unit wage and capital rental costs depicted in Figure 4.11. Thus suppose the firm chooses point L on the diagram. That would imply current unit wage costs given by the distance OK and unit capital costs are measured by the distance ON. The latter construction is achieved with NM parallel to BC. As point L is moved along the isotech it

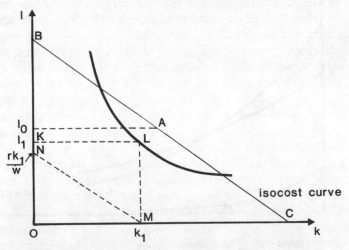

Figure 4.11: Choice of a new technique from the isotech

can be seen that higher unit capital costs imply lower unit wage cost, and vice versa.

Although the current purchase price of capital and the rate of interest at which it may be financed are not affected by the firm's expectations about the future price of its product, expected discounted quasi-rents do depend on such expectations. With a given discount rate ρ and expected rate of price decline α, the discounted price at any future time is given by:

$$p_t = p_0.\exp(-(\alpha + \rho)t)$$

Similarly discounted future unit labour costs are given by $wl_1.\exp(-\rho t)$ where l_1 is the unit labour requirement chosen from the range of possibilities given by this period's isotech.

When the price index of capital assets, denoted p_k, is constant and depreciation is assumed away so that the life-time of capital goods is infinite, the service price of capital is simply $p_k r$ where r is the interest rate at which the capital is financed.

If in addition it is assumed that the discount rate ρ is equal to r, the discounted service price [7] at time t is $p_k\rho.\exp(-\rho t)$, and the firm's choice problem of diagram 4.11 can be transposed to Figure 4.12.

The firm chooses points B, K, N such that the difference between

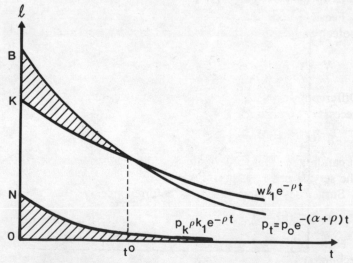

Figure 4.12: Determination of the rate of obsolescence

the hatched areas on the figure is maximised. The point B is fixed by the current price level, but K and N are open to choice within the constraints set by the current isotech and correspond to similarly labelled points on Figure 4.11. The point t^0 on the time axis represents the time at which the capital equipment installed in the current period becomes obsolete – the discounted quasi-rent at any point in time is given by the vertical difference between the curves emanating from points B and K on the vertical axis. The equipment will be scrapped at time t^0. It is assumed that scrapping costs are zero and that scrapped equipment has no resale value.

Since past 'vintages' of capital equipment may be used concurrently the analysis is directed towards discovering the implications about labour/capital ratios or rates of factor augmentation for increments in output, rather then for output as a whole.

The scene is set initially by assuming, in line with the analysis of the previous section, that firms suppose the current price of output to hold indefinitely over the future. The firm's choice problem is to choose b, the rate of labour augmentation in the current period, such that the present value of future unit operating surpluses minus the unit purchase cost of capital is maximised.

$$\max_b V = \max_b \left\{ \int_0^\infty (p_0 - wl_1)e^{-\rho t}dt - p_k k_1 \right\}$$

Feasible combinations of l_1 and k_1 are given by the current isotech and depend upon initial unit factor requirements, l_0 and k_0:

$$V = \int_0^\infty (p_0 - wl_0 e^{-b})e^{-\rho t}\,dt - p_k k_0 e^{-g(b)}$$

Differentiating with respect to b and equating to zero gives the necessary condition:

$$g'(b) = -wl_1/\rho p_k k_1 \qquad\qquad 4.16$$

It can be seen that this equation is identical to equation 4.11, since the service price of capital, $r = \rho.p_k$.

Sufficiency is guaranteed, as before, since:

$$p_k k_0 g''(b) < 0$$

Now, taking into account expectations of future advances in technology, assume the firm to expect a constant rate of price decline α in the future.

Then expected future price at time t is given by:

$$p(t) = p_0.\exp(-\alpha t), \quad \alpha > 0$$

where the initial time-period is indexed zero.

With a decision on unit labour requirements in the current period, l_1, the operating surplus at time t may be written:

$$s(t) = p(t) - wl_1$$

$$= p_0.\exp(-\alpha t) - wl_0.\exp(-b)$$

Therefore, the time at which such operating surpluses have fallen to zero (t^0) can be found by equating this expression to zero:

$$s(t^0) = 0 = p_0.\exp(-\alpha t) - wl_0.\exp(-b)$$

This solves for t^0:

$$t^0 = -(b/\alpha).\ln(wl_0/p_0)$$

Defining l/t^0 as the rate of obsolescence, it can be seen that the rate of obsolescence is a decreasing function of b:

$$d(l/t^0)/db = d(l/t^0)/dt^0.dt^0/db = \ln(wl_0/p_0)/\alpha t^{02} < 0$$

since $wl_0 < p_0$.

In other words the greater the increase in labour augmentation chosen (or the lower current unit labour requirements), the longer the equipment embodying these labour requirements will last with a given expected rate of price fall.

In terms of Figure 4.12, as point K moves downwards, so t^0 must move to the right. However, recall from Figure 4.11 that as point K moves downwards so point N must move upwards. The benefits of an increase in the rate of labour-augmentation in any period then are lower unit wage costs in future periods until obsolescence and postponement of the date by which that period's capital investment becomes obsolete. Offsetting these is the increase in capital costs in all future periods.

The firm's problem is then to choose b such that

$$V = \int_0^{t^0} [p_0 e^{-\delta t} - wl_0 e^{-b}].e^{-\rho t}dt - p_k k_0 e^{-g(b)}$$

is maximised.

Differentiation[8] gives:

$$V'(b) = t^{0'}(p_0 e^{-\alpha t} - wl_0 e^{-b})e^{-\rho t^0} + \int_0^{t^0} wl_0 e^{-(b+\rho t)} dt + g' p_k k_0 e^{-g}$$

Equating to zero, and noting that by definition of t^0 the first term in brackets vanishes,

$$g'(b) = -(wl_0e^{-b}/\rho p_k k_0 e^{-g})(1 - e^{-\rho t^0}) \qquad 4.17$$

It can be seen that:

$$\lim_{\alpha \to 0} g'(b) = -wl_1/\rho p_k k_1$$

which replicates the case analysed previously in which the current price is expected not to change in the future (equation 4.16).

Assuming that the sufficient conditions for a maximum are satisfied at the solution implied by equation 4.17, it can be inferred from that equation that the equilibrium rate of capital augmentation is greater than it would have been had no decline in output price been anticipated. This is so because the faster rate of obsolescence of capital implied by the falling output price means that the user cost of capital increases relative to the wage rate. This induces firms to choose a higher rate of capital augmentation at the cost of lower labour augmentation.

4.7   CONCLUSION

This chapter has examined the influence of factor prices on the derived demand for invention, and how they might induce a bias in the characteristics of invention towards productive techniques that economise on the 'scarce' factors. The theory was presented in section 4.2 in terms of cost functions, and a short review of the literature on empirical evidence of induced bias was presented in section 4.3. The evidence from the empirical studies is broadly consistent with the hypothesis that inventions are induced which save the scarce factor(s) of production. In other words, the invention process in the large seems to present a trade-off between labour-saving and capital-saving invention, and the remaining sections of the chapter explored the implications of such a trade-off, which is known as an invention possibility frontier. It is argued in section 4.5 that the invention possibility frontier will tend to induce Hicks-neutral technical change in the long run, which is a result that appears to contradict the growth theory idea of long-run Harrod-neutrality. The difference may be due to the fact that here, in a microeconomic context, factor prices are assumed exogenous whereas it is factor supplies that are exogenous at the level of macroeconomics. When productive firms have sunk costs in the shape of previous vintages of installed capital, and recognise that technical change reduces output price, there will be a tendency for them to opt for capital-augmenting new technologies; this is demonstrated in section 4.6.

NOTES

1. The apparent fact that capital has increased more rapidly than labour over the last century or so may, as Solow (1958) pointed out, be a statistical illusion: the result of measuring labour in man-hours instead of efficiency units.
2. Hicks states that:

   Substitution, in the sense in which we are using it, may take any of three forms: . . .
   (3) The changed relative prices will stimulate the research for new methods of production which will use more of the now cheaper factor and less of the expensive one. Partly, therefore, substitution takes place by a change in the proportions in which productive resources are distributed among existing types of production. But partly it takes place by affording a stimulus to the invention of new types. We cannot really separate, in consequence, our analysis of the effects of changes in the supply of capital and labour from our analysis of the effects of invention. (op. cit., p. 120)

3. Binswanger argues that, given that innovation possibilities are greater for machinery-using methods, the synchronised swings in the bias and relative price which he observed, with the former lagging, support the induced innovation hypothesis.
4. Edwin Burmeister and Rodney Dobell (1969) prove that Harrod-neutrality implies this particular form from the definition of Harrod-neutrality, that 'when all output—capital ratios . . . remain constant, so also do all capital rentals . . . (and hence all profit shares).
5. William Nordhaus (1969) has argued that the *deux ex machina* nature of the function may be needed to preserve competition, because if the IPF can be shifted by devoting scarce resources to do so, there would appear to be increasing returns to scale overall if there are constant returns to 'normal' factor inputs.
6. Denote by p_s the service price of capital, and by p_k its asset (purchase) price, where $\dot{p}_k$ is its rate of change. Let r be the supply price of finance (the interest rate) and δ the rate of depreciation, which includes obsolescence. Then the price of capital services in a competitive hiring market determines p_s, which is given by:

   $$p_s = p_k(r + \delta - \dot{p}_k)$$

7. Note that the asset price is recovered as the discounted sum of service prices:

$$p_k = \int_0^\infty p_k . \rho . e^{-\rho t} dt$$

8. Using the rule that if $Y = \int_{T_1(x)}^{T_2(x)} F(x, T) \, dt$ and T_1, T_2, F possesses continuous derivatives then:

$$Y' = T_1' . F(x, T_1(x)) + T_2' . F(x, T_2(x)) + \int_{T_1(x)}^{T_2(x)} F_x(x, t) \, dt.$$

5 Invention Market Organisation

This chapter analyses competition between inventors. The assumption that competition is beneficial for the allocation of resources cannot be maintained in this context. Section 5.1 outlines the problems of an invention market and serves as a motivating introduction to the subsequent sections of the chapter, which model the inventive process according to different paradigms. One such paradigm is that of an open access pool of resources. Another is that of a race to be first. These are examined in sections 5.2–5.4, in each of which comparisons are made between the case of a lone inventor and that of an inventor in a competitive environment with a large number of competitors. Finally, in section 5.5, the large numbers assumption is dropped and the inventor is assumed to compete against a small number of interacting opponents.

5.1 THE PROBLEMS OF AN INVENTION MARKET

The economic characteristics of the inventive process and of inventive output preclude the possibility of efficient resource allocation in this area by an unregulated market. The problem characteristics are partly those shared by any activity for which information is an output. Broadcasting is an example. The essential point is that information, which is costly to acquire, can be passed on at very low costs. And the costs of supplying information, once produced, do not depend on its value, nor on the number of recipients. Economic efficiency therefore requires that, with a given amount of information in existence, it should be available at zero price. But this does not solve the problem of how much information to produce in the first place. If there were to be no reward it would undoubtedly be produced in too small volume. It has been claimed that the rewards should be separated from the (zero) price. But then there is no mechanism to produce either the right amount or the right type of information. The social solution to this problem is the institution of patent rights, which allow the inventor to extract a proportion of the economic value of the invention.

In this chapter, as in others, the institution of patent rights is taken as given. They exist in conjunction with other devices, such as secrecy and monopolising of production, to extract private benefit from information production. But though patents and other ways of appropriating the social benefits of inventive activity imply some sort of monopoly in the *use* of inventions, they are quite consistent with competitive production of inventions. The problem with a monopoly in the use of inventions is that inventions will cooperate with too few resources in production; but the problem with a competitive supply of inventions is that too many resources will be attracted to their production.

An obvious drawback of competition in the production of inventions is that one team of researchers may simply duplicate the work of another team. But even if this duplication could somehow be avoided, there is another, more subtle disadvantage in the competitive supply of inventions which arises from the fact that researchers are fishing for results from a common pool of problems. This pool is itself a resource, which is manifested by the fact that there are richer pickings for applied research leading to inventions when science throws up a larger number of possible problems or approaches than when it has little to offer. As far as the invention market is concerned, the economically relevant characteristic of science is the amount of rent that the realisation of its potential can give rise to.

Two facets of the common pool problem that afflicts the allocation of resources to inventive activity are adumbrated in sections 5.2 and 5.3. The problem arises because of the impossibility, due to social arrangements, of appropriating the benefits derived from a scarce common resource. In the present context the common resource is the state of science and technology or, more precisely, the as yet untapped inventions derivable from science and technology.

The competitive behaviour of inventors is often modelled as a race to be first. An important conclusion of such analysis is the proposition that a race with more than one participant will result in 'too early' invention compared with the social optimum, whereas monopolistically supplied invention services will be provided with optimal timing. A formal equivalence exists[1] between these competitive race models and the problem of open access to a common pool of resources: the characteristic feature of both is the dissipation of economic rent through the competitive process. In fact, it is the pool of unappropriable economic rent that is subject to 'the rule of capture' in both cases.

The essential nature of the problem is outlined in section 5.2 as a 'fishing problem', and then more explicitly and in more detail in section 5.3 as a race to be first. Normative implications of the race to be first are taken up in section 5.4, where it is assumed that variations in the term of patent protection and of the royalty that an inventor can charge are instruments available to a social planner attempting to achieve an efficient allocation of resources. In section 5.5 the race element in technological competition is ignored so as to focus on the interplay of small-number rivalry and rent-seeking as determining elements of the waste of resources inherent in technological competition.

5.2   RESEARCH AS A FISHING PROCESS

Consider research inputs as homogeneous composites,[2] so that it is possible to speak of a unit of research resources. This simplifies matters, though it does not affect the substance of the argument that follows, which could instead be cast in terms of a vector of heterogeneous research inputs. For concreteness it is helpful occasionally to refer to a unit of research resources as 'a researcher'. Research inputs are supplied to the inventive sector as an increasing function of the rental rate at which their services are hired.[3] That is, the supply curve of researchers is upward-sloping in their factor price. At any moment of time there are numerous fields of research being pursued which, like land, vary in fertility and productiveness. Within each field researchers pursue distinct projects, chosen rationally according to *ex-ante* beliefs about profitability with full awareness of any overlaps or duplications with other research projects. They therefore position their choice of project in the relevant characteristics space for research output so as to maximise expected net revenue, and in so doing ensure that each project earns equal net revenue in an expected value sense, it being assumed that researchers can appropriate the full social value of any inventions they produce. Researchers are like fishermen[4] working a fishing ground and, pursuing the analogy, the research project is equivalent to the choice of location from which to cast a net while inventions correspond to the catch of fish. Assuming equal competence, effort and equipment among fishermen, it is clear that each would have the same expectation of landings and of profit.

If research resources are free to migrate between fields of application, and if there is open access to any such field, then migration will indeed take place so long as expected average net revenue differs between fields. Migration occurs until the expected average net revenue is equalised between fields. If fields of application are small relative to the inventive sector as a whole, then within each field marginal and average cost of the variable resources are constant and equal, so that in equilibrium average value product is equalised between fields. At the extensive margin the net profitability of a research field is zero, so by the foregoing argument it must also be zero in intramarginal fields because of the competition of research resources at the intensive margin within each. It follows that open access to research fields combined with the competitive mobility of researchers between fields has driven rents to zero.

In contrast to the resource allocation achieved under open access to research fields, if each field of research were monopsonised in some way, implying blocked entry, the various monopsonists would hire research inputs up to the point that marginal cost equals marginal value product. Such an allocation implies less research inputs in each field of research than under free access because the supply of researchers is increasing in researcher wages, so that marginal cost is a rising function of research output. It also implies that rents are maximised.

5.3 RESEARCH AS A RACE TO BE FIRST

The requirement of precedence in patenting inventions clearly implies a race between research organisations whose activities are leading to the same or equivalent technical specifications in the invention. This is self-evident. It is also clear that a race exists between research teams aiming for substitute inventions, even where the technical specifications differ so that proof of precedence for patent purposes is not at issue. But it was not widely recognised until Yoram Barzel's (1968) contribution that such competitive races to be first involve a waste of resources.[5] They imply an over-supply of inventive inputs compared to the social optimum, and invention takes place too early as a result. This is once more a reflection of the common pool problem because the characteristic feature of the race to be first is erosion of the quasi-rents which represent the social value of the invention. The fundamental cause of rent dissipation here, as before, is the overcrowding of science by inventors.

Time is an essential feature of a race to be first. This fact points to the relevance of the time–cost tradeoff in research projects outlined in Chapter 2. This tradeoff is utilised in the analysis that follows. The basic results derived by Barzel did not allow for the tradeoff: he assumed that the time elasticity of supply of the innovation was zero. Recognition of the fact that extra resources can speed up an invention modifies, but does not overturn, Barzel's basic insight.

It is assumed that a particular invention is sought. The cases to distinguish are first that in which the inventor knows that he is the only seeker, and secondly when he believes himself to be in a race with an unknown number of unidentified competitors. These cases will be compared with the allocation of resources over time that would be decided in a socially managed economy by an omniscient resource allocator.

Granted that the inventor faces a competitive invention-using industry and can appropriate the fruits of his minor cost-reducing invention through an efficient and indefinite patent monopoly, the basic analysis of Chapter 3 implies that his royalty revenue in any post-invention period is given by the output level of the industry in which the invention is used, multiplied by the reduction in unit costs. Apart from the deadweight loss implicit in the patent monopoly, this royalty revenue equals the rate of social benefit from the invention. Assume that royalty revenue R^0 is constant each post-invention period; then its present value at any time T years prior to completion of the research is given by:

$$R(T) = \int_T^\infty R^0 . e^{-\rho t} \, dt = R^0 . e^{-\rho t}/\rho$$

where $\rho > 0$ is some suitable rate of discount. Differentiating:

$$R'(T) = -R^0 e^{-\rho T} < 0$$

and

$$R''(T) = \rho . R^0 e^{-\rho T} > 0.$$

The function is shown in Figure 5.1.

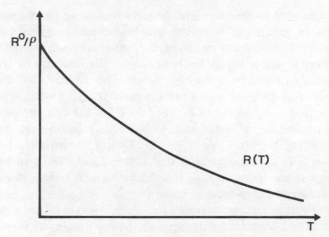

Figure 5.1: Inventor's discounted revenue function

On the cost side, use is now made of the time–cost trade-off discussed in Chapter 2. This implies that the (discounted) cost of a research project is a decreasing function of the remaining time required to complete the project. It is also usually thought of as a convex function, as in Figure 5.2.

In the absence of rivalry from other inventors, the profit-maximising inventor will choose completion time T such that the net present value of his activity is maximised – he chooses T so as to maximise the difference between R(T) and C(T).

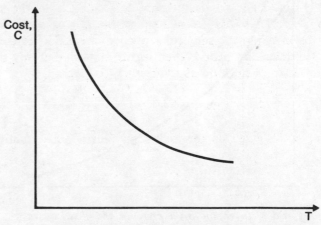

Diagram 5.2: Inventor's time–cost trade-off

By contrast, an inventor who believes himself to be in a race to be first to patent the invention will be inclined to speed up the research process thereby incurring extra costs according to the C(T) schedule. In the limit, as his belief about the intensity of rivalry increases, all profit from the invention will be eroded away to leave him with a barely positive net present value from inventive activity. The situation is depicted in Figure 5.3. T^m is the completion time for the monopoly inventor and T^c is the completion time for the competitive inventor. At T^c, $C(T) = R(T)$ whereas at T^m, $C'(T) = R'(T)$. It can be seen that competition among inventors forces a more intensive R & D programme and results in earlier completion of the project.

Competition among inventors therefore involves higher costs of producing an invention compared to the monopoly inventor case on two counts. It involves duplication of costs between inventors and also for any one inventor it forces higher costs due to earlier completion than would otherwise be the case.

It may be noted that for any completion time T to be feasible (i.e. not to involve the inventor in losses), the curve C must lie on or below the curve R at that T, i.e. $C(T) \leq R(T)$. However, it is not necessary for C(T) to be globally convex. Local convexity of $\{R(T) - C(T)\}$ is sufficient for the conclusion that potential competition results in each inventor incurring higher costs and intro-

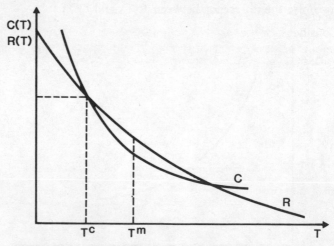

Figure 5.3: Invention in the presence and absence rivalry

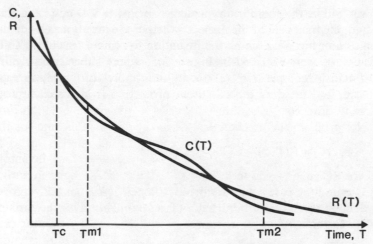

Figure 5.4: Local non-convexity in the cost–time trade-off

ducing the invention earlier than in its absence. So long as the invention is feasible, ie. $C(T) \leq R(T)$, the conclusion stands. This is illustrated in Figure 5.4. Here, the monopolist inventor chooses a timescale for the R & D project, such that T equals either T^{m1} or T^{m2}, depending upon which gives the larger profits. The competitive inventor is still forced to choose $T = T^{c}$ as before. In either case the inventor's costs will be higher if he is faced with potential competition in the production of the invention.

Consider now how a finite period of patent protection affects the foregoing analysis. The cost function is not affected, but the inventor's discounted revenue function is now given by:

$$R(T, \theta) = \int_{T}^{T + \theta} R^{0} e^{-\rho t}\, dt = (1 - e^{-\rho t}) R^{0} e^{-\rho T}/\rho \qquad 5.1$$

where θ is the lifetime of the patent. The slope of the revenue function is:

$$R_{T}(T, \theta) = -(1 - e^{-\rho \theta}).R^{0}.e^{-\rho T}$$

With a finite period, $\theta < \infty$, implying $e^{-\rho \theta} > 0$, it follows that:

$$R(T, \theta) < R(T, \infty) \text{ and } R_{T}(T, \theta) > R_{T}(T, \infty)$$

where the function $R(T, \infty)$ corresponds to the previous $R(T)$.

Furthermore, the shorter the period of patent protection, the lower will be the equilibrium present value of R & D costs and the longer the timescale of the project, whether the inventor faces competition or not – assuming the invention to remain feasible. These results are demonstrated in Figure 5.5, where superscripts on T distinguish 'competitive' (c) from 'monopoly' (m) invention as before, and the duration of patent protection is indicated by the term in brackets.

Net profit to the inventor is:

$$\pi(T, \theta) = R(T, \theta) - C(T)$$

where $R(T, \theta)$ is given in 5.1.

The limiting case under rivalrous competition among inventors is defined by the condition that $\pi(T, \theta) = 0$. This gives the timing of invention as:

$$T^c(\theta) = - \ln\{\rho C/R^0(1 - e^{-\rho\theta})\}/\rho$$

In the absence of rivalry, the timing of the invention is given from maximisation of $\pi(T, \theta)$ with respect to T for any patent duration θ:

$$T^m(\theta) = - \ln\{- C_T/R^0(1 - e^{-\rho\theta})\}/\rho \qquad 5.2$$

and the difference $T^c - T^m = - \ln\{- \rho C/C_T\}/\rho$

The comparative static propositions implicit in Figure 5.5 are

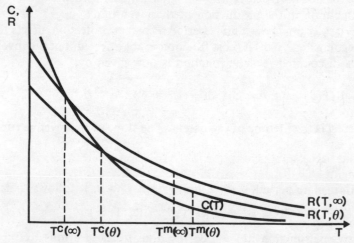

Figure 5.5: Effect of a finite period of patent protection

established by signing the derivative $dT/d\theta$ implicit in the relevant equilibrium conditions. For the competitive case the equilibrium conditions are:

$$R(T, \theta) - C(T) = 0 \text{ and } C_T(T) < R_T(T, \theta).$$

Taking the total differential of the equality with respect to T and θ gives the derivative:

$$dT^c/d\theta = R_\theta/(C_T - R_T)$$

in which the denominator is negative (from the equilibrium conditions), and the numerator is positive since:

$$R_\theta = \partial R/\partial\theta = R^0 e^{-\rho(\theta + T)} > 0$$

hence:

$$dT^c/d\theta < 0.$$

In the monopoly case the equilibrium conditions are:

$$R_T - C_T = 0 \text{ and } R_{TT} - C_{TT} < 0$$

Again, taking the total differential of the equality, with respect to T and θ gives:

$$dT/d\theta = R_{T\theta}/(C_{TT} - R_{TT})$$

From the equilibrium conditions, the denominator is positive. The numerator is:

$$R_{T\theta} \equiv \partial^2 R/\partial\theta^2 = -R^0 p e^{-\rho(T+\theta)} < 0$$

hence:

$$dT^m/d\theta < 0.$$

5.4   NORMATIVE IMPLICATIONS OF THE RACE TO BE FIRST

It has been established so far that when invention takes place under imagined rivalry in a race to be first, the inventor will devote more resources to research and achieve the invention in a shorter time than he would have done in the believed absence of rivalry. To avoid the complications of duplicated research projects in the case of rivalrous invention, suppose that the speed and intensity of inventive activity must be publicly declared before the resources are

actually committed, as if determined in an auction under sealed
bids. The question now is: though invention under conditions of
rivalry is faster than in their absence, is it more or less socially
desirable? Yoram Barzel (1968) had concluded that a single
monopolistic innovator would organise the timing of an innovation
to maximise his quasi-rents and thereby also maximise social
welfare. In the present context, however, there is a deadweight loss
of social welfare due to the output-restricting effect of the post-
invention patent monopoly. It will now be shown that this
deadweight social loss implies that, in the absence of rivalry, invention takes too leisurely a pace, even when the duration of patent
protection is permanent. But, on the other hand, invention under
imagined rivalry is, in the limit as all rents are eroded away, too
fast, as Barzal argued.

The social value of an invention is given by the discounted sum
of increments to producers' and consumers' surpluses that it
enables. The undiscounted annual rate of social value from a minor
cost-reducing invention is shown by the shaded quadrilateral of
Figure 5.6. While the patent is protecting the inventor the social
gain per period is given by the shaded rectangle R^0, which equals
the inventor's maximal royalty revenue. After the patent protection
has lapsed the whole shaded quadrilateral represents social gain in

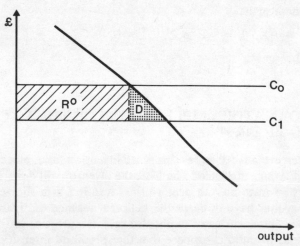

Figure 5.6: Instantaneous increment to social value

the form of consumers' surplus. The triangle D represents deadweight social loss during the period of patent protection.

The socially optimal duration of patent protection would be such as to equate the marginal gain in consumers' surplus to the marginal social loss caused by decreased inventive activity as the period of patent protection is reduced. The problem confronting a social planner is therefore to set the duration of patent protection to maximise net social gains subject to the behaviour of inventors.[6] He must choose both the speed of invention T and the duration of patent protection θ to maximise social welfare:

$$\text{Max } W(T, \theta) = \int_T^\infty R^0 e^{-\rho t}\, dt + \int_{T+\theta}^\infty D e^{-\rho t}\, dt - C(T) \qquad 5.3$$

subject to either (a) $R(T) - C(T) = 0$ for invention under rivalry or (b) $R_T(T, \theta) - C_T(T) = 0$ in the absence of rivalry.

With rivalrous invention, the Lagrangean for this problem is:

$$R^0 e^{-\rho T}/\rho + D e^{-\rho(T+\theta)}/\rho - C(T) + \lambda\{(1 - e^{-\rho\theta})R^0 e^{-\rho T}/\rho - C(T)\},$$

which yields the three first-order conditions:

$$- R^0 e^{-\rho T} - D e^{-\rho(T+\theta)} - C_T - \lambda\{(1 - e^{-\rho\theta})R^0 e^{-\rho T} + C_T\} = 0$$
$$- D e^{-\rho(T+\theta)} + \lambda R^0 e^{-\rho(T+\theta)} = 0$$
$$(1 - e^{-\rho\theta})R^0 e^{-\rho T}/\rho - C = 0$$

The Lagrange multiplier represents the marginal social valuation of an increment in the profitability of invention. From the condition that $\partial L/\partial \theta = 0$ it can be seen that $\lambda = D/R^0$ and hence, on substitution and arranging:

$$T^* = - \ln(- C_T/R^0)/\rho \qquad 5.4$$

which is achieved by setting the period of patent protection to:

$$\theta^* = - \ln(1 + \rho C/C_T)/\rho$$

The interpretation of this solution is interesting: it implies that the period of patent protection must be set so as to induce the rivalrous invention at the same time as an unrivalled monopoly inventor with indefinite patent protection would produce it. This can be seen by comparing 5.4 with 5.2.

Consider now the socially correct period of patent protection for inventions produced in the absence of rivalry. Now the problem is that of welfare maximisation as in 5.3 with constraint (b). For this the Lagrangean is:

$$R^0 e^{-\rho T}/\rho + De^{-\rho(T+\theta)}/\rho - C(T) + \lambda\{R^0 e^{-\rho T}(1 - e^{-\rho\theta}) + C_T(T)\}$$

for which the three first-order conditions are:

$$-R^0 e^{-\rho T} - De^{-\rho(T+\theta)} - C_T - \lambda R^0 \rho e^{-\rho T}(1 - e^{-\rho\theta}) + \lambda C_{TT} = 0$$
$$- De^{-\rho(T+\theta)} + \lambda\rho R^0 e^{-\rho(T+\theta)} = 0$$
$$R^0 e^{-\rho T}(1 - e^{-\rho\theta}) + C_T = 0$$

So that $\lambda = D/\rho R^0$ and, on substitution and arranging:

$$T^{**} = -\ln\{[(DC_{TT}/\rho R^0) - C_T]/(R^0 + D)\}/\rho \qquad 5.5$$

and:

$$\theta^{**} = -\ln\{(C_{TT} + \rho C_T)/(C_{TT} - \rho C_T R^0/D)\}/\rho \qquad 5.6$$

Note that with totally inelastic demand for the final product the deadweight loss triangle in Figure 5.6 disappears and $D = 0$, so the optimal timing of the invention is once more that which an unrivalled monopolist would introduce if he had permanent patent protection. This can be seen by substituting $D = 0$ into 5.5 and comparing with 5.2. The same substitution into 5.6 would make the denominator in brackets infinitely large, and since the logarithm of zero is minus infinity, permanent patent protection would be implicit ($\theta^{**} \to \infty$). But any non-zero elasticity of demand would imply that the Marshallian triangle measured by D is positive, and hence that patent protection in this case should be less than permanent. Indeed, since the term in braces in equation 5.6 must lie between zero and one,[7] it follows that the optimal duration of patent protection is positive and finite for a non-vertical demand curve.

Pankaj Tandon (1982) also considered the question of optimal patents, but introduced a new degree of freedom for the social planner – namely, the royalty rate. Tandon assumes that patents are privileges granted on the condition that the patentee licenses out the technology at a socially optimal royalty. In other respects his model is similar to that of Nordhaus (1969) in its representation of the invention production function and in considering only the non-rivalrous case. Tandon concludes that social welfare is more sensitive to changes in the (socially optimal) royalty rate than to changes in patent life. In fact, when the planner can fix both the royalty and the life of the patent it is optimal to set an infinite life. He also demonstrates, for the case of new product invention which is equivalent to drastic cost-reducing invention, that with a given finite patent life the optimal royalty is a decreasing function of patent life, of the discount rate and of the curvature of the invention production function. But it is not very sensitive to any of these.

To examine the effect of a socially optimally determined royalty rate with rivalrous invention, consider Figure 5.7. The inventor receives $R(r)$ for the duration of patent protection while consumers' surplus of $B(r)$ exists for all time and is increased to $B(r) + R(r) + D(r)$ after the patent has expired where $D(r)$ is the deadweight cost of the patent monopoly for the period of its duration. The problem is to maximise discounted social welfare, given by:

$$W(r, \theta, T) = (B(r) + R(r))e^{-\rho T}/\rho + D(r)e^{-\rho(T + \theta)}/\rho - C(T)$$

subject to the zero discounted profit condition:

$$(1 - e^{-\rho\theta}).R(r)e^{-\rho T}/\rho - C(T) = 0$$

Forming the Lagrangean, and equating its partial derivative with respect to r to zero gives:

$$\{(B_r + R_r)e^{-\rho T}/\rho + D_r e^{-\rho(T + \theta)} + \lambda R_r.(1 - e^{-\rho\theta})e^{-\rho T}\}/\rho = 0$$

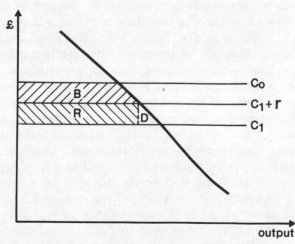

Figure 5.7: Socially optimal royalty rate

which implies one of the following: $T = \infty$ (no invention); $\theta = 0$ (again, no invention), or $\lambda = D_r/R_r$, in view of the fact that $R_r + B_r = -D_r$. Only the last of these is economically interesting, but in turn it is only consistent with the other first-order conditions if the period of patent protection θ is infinite. Making this assumption, it then follows that the invention is introduced at time T given by:

$$T = -\ln\{-C_T(D_r + R_r)/R(D_r + R_r + R_rB/R)\}/\rho \qquad 5.7$$

Comparing 5.7 with 5.4, it is clear that the socially optimal timing of invention under rivalry when the social planner additionally controls the royalty rate is slower then when he only determines the duration of patent protection.

5.5 STRATEGIC FACTORS IN COMPETITIVE INVENTION

It was assumed in section 5.3 that competition among inventors would drive away all potential profit from the research activities. The competitive equilibrium was characterised by zero profits. This is usually achieved in economic theory by a combination of two assumptions: large numbers of competing firms, and freedom of entry and exit into the industry. It might be asked whether these assumptions are reasonable ones to make in the case of technological competition. The idea that there might be a large number of inventors trying to produce precisely the same invention hardly seems realistic. But, as the theory of contestable markets has shown, [8] the large numbers assumption is not actually necessary to derive a zero profit equilibrium. Freedom of entry and of exit, together with absence of sunk costs, are sufficient to guarantee that a monopoly, even a natural monopoly, will behave like a competitive industry. It does so in the fear of a 'smash and grab' raid on any super-normal profit by potential entrants. This might not be a totally implausible description of the conditions in which competing research laboratories operate. As such it might perhaps serve as the paradigm under which the analysis of section 5.3 holds. In this section, however, it will be assumed that the typical inventor or research laboratory knows itself to be in an industry with a limited number of competitors. In these circumstances the competitive nature of its activities must be modelled by taking into account strategic factors.

The question how would competing inventors behave when they recognise the interdependence between their decisions and those of other inventors is what makes the problem 'strategic'. The interdependence between decisions implies that the inventive firms cannot simply be modelled as reacting to changing parameters in their environment. The problem is the familiar one of oligopoly. There is, of course, a large variety of possible approaches, depending on how the conjectural variations of the players are modelled. In any event, it seems that recourse to game theory is needed to deal with the situation. And a simple yet reasonable formulation would be to assume that the competitors have Cournot–Nash conjectural variations. That is to say, each competitor believes that there will be no retaliation to its own decisions: it can take its rival's actions as given. The Cournot–Nash equilibrium is the intersection of all participants' reaction functions. Usually, as the number of competitors increases, the Cournot–Nash equilibrium converges on that of pure competition.

Utilising the concept of a time–cost trade-off in the context of an oligopoly of researching firms, all entertaining Cournot conjectures, Partha Dasgupta and Joseph Stiglitz (1980) have shown that 'with free entry, at most one firm will be engaged in R & D activity at an equilibrium, and its net present value of profits will be nil'. They achieve this result assuming absence of uncertainty – the same assumption as has been used in section 5.3 and 5.4 above. This can therefore justify the ignoring of duplication as a cost of competition in races to be first when the time–cost trade-off is known with certainty.[9] An implication of Dasgupta and Stiglitz's result is that the desirability of monopoly that is Barzel's basic insight, discussed in 5.4, means *ex-ante* monopoly: the researching firm must know before hand that it is the only researcher for it to produce the optimal timing of an invention.

The foregoing has implied that duplication of research work is not a necessary aspect of the social wastes of technological competition. But this is an illusion. There is a rather subtle form of duplication taking place in the contraction of research time along the time–cost trade-off. It is exemplified by the paradigm of parallel and sequential research efforts presented in section 2.4. In the remainder of this section the time–cost trade-off is eschewed in favour of a simpler fishing paradigm of technological competition, which makes clear the connection between the social cost of free access to a common property resource and the wastes of duplication. The common property resource is the potential solution to a technical problem which can be patented.

The undesirable consequences of competition among inventors for the monopoly right of a patent were noted by Arnold Plant (1934). He observed that, unlike rights over 'normal' property, the property right in a patent is not a consequence of scarcity, but rather it makes possible the creation of a scarcity which could otherwise not be maintained. Of course, the objective of the patent is to encourage the activity of invention. But, Plant points out, this objective is only achieved if the activity of invention actually responds to economic inducements: which is obviously not the case for what Hicks (1932) terms 'autonomous inventions'. And even if inventions can be induced by rewards, there is a further problem with the institution of patent monopolies established by legal precedence: namely, that excessive resources are drawn to the production of patentable inventions. Plant takes to task those economists who, like J. B. Clark, F. W. Taussig and Jeremy Bentham, defend the patent system but do not take into account the wastes of this technological competition.[10] As Plant says, '[T]he effect must surely be to induce a considerable volume of activity to be diverted from other spheres to the attempt to make inventions of a patentable type' (op. cit., p. 42).

The problem that Sir Arnold Plant recognised half a century ago can be seen as an example of a wider class of competitive activities that are socially wasteful. In recent years they have acquired the names 'rent-seeking' (Buchanan *et al.*, 1980) and 'DUP' (Bhagwati, 1982) activities.[11] The common characteristic is that the competition for an artificially created scarcity absorbs genuinely scarce resources. Examples abound where public authorities issue licences, or otherwise regulate an industry. In these cases the social cost of monopoly is not simply the deadweight loss of the Marshall–Harberger triangle (BCE in Figure 5.8), but can even absorb the whole rectangle to its left, the hatched area shown in Figure 5.8.

Thus Richard Posner (1975) argues that competition to obtain a monopoly would be carried out to the point where, at the margin, the cost of obtaining the monopoly is equal to the expected benefit

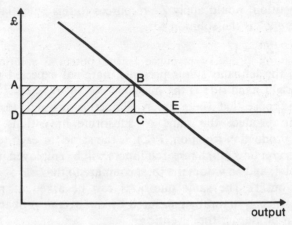

Figure 5.8: The social cost of rent-seeking

of being a monopolist. He assumes that the long-run supply of all inputs used in obtaining such monopolies is perfectly elastic. This ensures that 'all expected monopoly rents are transformed into social costs', i.e. that the hatched area in Figure 5.8 is absorbed as a cost. But even where the supply of inputs is inelastic, he suggests that 'in the long run the availability of such rents will attract additional resources into the production of those inputs, and these resources will be wasted from a social standpoint'.

Consider a simple fishing model of research in which a particular invention is sought. Suppose that it would result in a reduction in unit cost in some industry of ξ^0 where it would be applied to output Q^0. Both ξ^0 and Q^0 are known to the researchers at the outset, and hence so is the potential revenue R^0 from the invention. If a research team devotes resources Z to the production of the invention it has a probability $P(Z)$ of success where $P(0) = 0$, $P' > 0$ and $P'' < 0$. These conditions reflect the basic notion of positive but diminishing marginal product to research activity. With this given there are two possibilities: first, that each research team has a production function of this sort, no matter how many research teams there are; secondly, that the research production function applies to the whole sector, and that the output of any research team is in proportion to the resource share of sectoral inputs that it employs. In either case, a single profit-maximising research team, untroubled by competition, would apply Z^* resources to this potential invention where Z^* is the solution to:

$$R^0.P'(Z^*) = 1$$

in which the lefthand side represents marginal expected revenue and the right hand side is the marginal cost of research.[12]

Now suppose that there are n competing research teams, all aiming to produce the same or substitute imventions. If the research production function, $P(Z)$, is the same for each team the questions are: how much research inputs will be employed by each; and in total; and how does the total compare to the case of a single research outfit? The same questions can be asked of research output, which is naturally measured by the overall probability of success in producing the invention.

Let Z_i be the research inputs employed by the ith team. Then the expected value of the research for the ith team, assuming no others were in the game, would be $R^0.P(Z_i)$. But even if the ith research team were successful in producing the invention there is no certainty that it would get the patent rights since other teams are also doing research on the problem. It is as if each team simultaneously enters a lottery to determine the allocation of the revenue R^0 from the invention, where the share of lottery tickets in possession of the ith team is equal to $P(Z_i)/\sum P(Z_j)$. Thus research purchases simultaneously a probability $P(Z_i)$ of producing the invention and a chance of getting the monopoly rent in the patent. The product of these gives the overall probability of extracting the rent in the potentially patentable invention. Hence the expected net value of research for the ith team is:

$$V_i = R^0.P(Z_i)^2/\sum P(Z_j) - Z_i \qquad 5.8$$

The symmetric Cournot–Nash equilibrium is obtained by maximising V_i with respect to Z_i for given $\{Z_j : j \neq i\}$, and noting that all Z_j $(j = 1,\ldots, n)$ will be equal $(= Z^{**}$, say). Each research team thus employs research resources satisfying:

$$R^0.P'(Z^{**}).[(2n - 1)/n^2] = 1 \qquad 5.9$$

The lefthand side of 5.9 is the expected marginal value product (EMVP) of research for any one of the research teams. It is a function of the number of research teams. The righthand side of 5.9 represents the marginal cost of research (normalised to unity). As

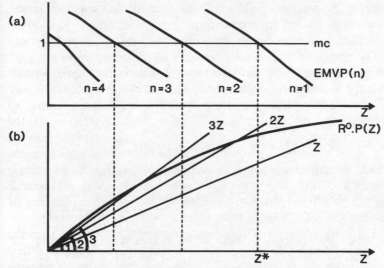

Figure 5.9: Size and number of research units with Cournot conjectures

the number of competing teams increases so the EMVP falls, and Z^{**} along with it. This is illustrated in part (a) of Figure 5.9. If there is free entry of research teams, then n increases so long as there is any positive expected profit; that is, so long as:

$$R^0.P(Z^{**})/n - Z^{**} > 0.$$

The area above the mc curve but below the EMVP curve in part (a) of Figure 5.9 represents expected profit for the representative research team. It is also depicted as the vertical difference between the $R^0.P(Z)$ curve and the n.Z ray in part (b) of the figure. So long as profit is positive, n increases and the EMVP curves shift downwards while the n.Z rays rotate anticlockwise until an extra entrant makes the representative research team unprofitable.

Total equilibrium research inputs in this model are $n.Z^{**}$. Since Z^{**}, the representative research team's input, is a declining function of n, total research inputs may either increase or decrease with n.[13] The deciding factor is the curvature of the research production function at the equilibrium for the representative research outfit, $P''(Z^{**})/P'(Z^{**})$. Another way of expressing this is to note that 5.9 implies a negative relationship between the equilibrium size of the representative research team, Z^{**}, and the number of research teams n. So long as the absolute elasticity of size with respect of numbers is greater then one, total research effort increases with increasing numbers of research teams.[14] But this is not guaranteed.

The output of the overall research effort is the probability of success in producting the invention. If there is no duplication of research between research teams, the individual probabilities of success are independent and the overall probability of producing the invention is:

$$1 - (1 - P(Z^{**}))^n = 1 - \exp(n.\ln(1 - P(Z^{**})))$$

which, on differentiation with respect to n, yields:

$$d(\ )/dn = -\ln(1 - P(Z^{**})).\exp(n \ln(1 - P(Z^{**})).$$
$$\{d(-P(Z^{**}))/dn\}$$

in which the expression before the braces is positive, and the derivative in braces, being $-(dP(Z^{**})/dZ^{**}).(dZ^{**}/dn)$, is also positive. Hence, the overall probability of success is an increasing function of the number of Cournot competitors.

On the other hand, with perfect duplication of research efforts, the overall probability of success is just the probability that the representative research team will produce the invention. And this must be a declining function of the number of competing teams, since optimal size of the representative research outfit falls as n rises. It follows that in the realistic case where some, but not all, of the research pursued by one team is duplicated elsewhere, the sign of the response of research output to the number of research teams depends on the extent of duplication.

In summary, within the confines of this simple model, the socially optimal number of Cournot-competing research teams for a given total of research resources need not be just one. Some duplication can be tolerated in exchange for the higher marginal productivity achieved in smaller teams.

Suppose now, however, that it is *not* assumed that the research production function applies to each team of researchers, but only to the research sector as a whole. Assume that the share of each research team in total output is in proportion to its share of research inputs. Now the optimisation problems for the ith team entertaining Cournot conjectures is to maximise:

$$V_i = R^0.P(\textstyle\sum Z_j).Z_i/\textstyle\sum Z_j - Z_i \qquad\qquad 5.10$$

Objective function 5.10 should be compared to the function 5.8 posited earlier. The term $P(\sum Z_j)$ represents the probability that the invention will be achieved by the sector taken as a whole, while $Z_i/\sum Z_j$ is the share of the ith team in total inputs. Maximising 5.10 with respect to Z_i and applying the symmetry of the Cournot assumptions yields the optimal input Z^{**} of the representative team for a given number of research teams, n, as the solution for Z of:

$$R^0.(P' + P/Z - P/nZ).n = 1 \qquad 5.11$$

Total research resources, $n.Z^{**}$, increase with increasing numbers of research teams since the total derivative $d(nZ^{**})/dn$ is:

$$-n.(P'' + (n - 1).(P' - P/nZ^{**})/Z^{**})/(P/nZ^{**} - P'),$$

which is positive. So too does output, $P(nZ^{**})$.

Now if it is assumed that entry takes place so long as expected profit is positive (i.e. so long as $R^0.P(nZ^{**}) > nZ^{**}$), then all the potential social benefits will be eroded away as costs, with the situation depicted in Figure 5.10. But to produce this solution all that is required is the threat of free entry. The entry need not actually take place. This free entry solution is therefore compatible with any number of research teams; the number is indeterminate. It is

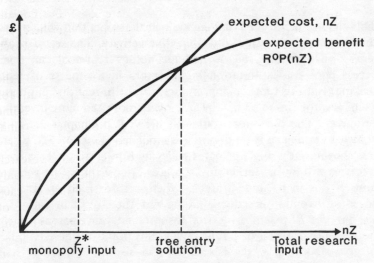

Figure 5.10: Monopoly and competitive free entry solutions

perfectly analogous to the 'common pool' solution which equates average cost and average expected benefits. Each pool of research problems is 'over-fished'. With research resources free to migrate between such pools, and being supplied with a finite price elasticity, the global equilibrium unit price of research resources is such as to equate the average expected benefit of research in all fields of research endeavour. But even the rent component of inelastically supplied researcher salaries will tend to be eroded by an excessive investment into education as resources are drawn into research from other sectors.

Note that duplication of research, in the normal sense of the term, was ruled out by the production function assumption that the research output of a sector is determined by the sum of inputs of all research teams. This form of externality, jointness in supply, makes a single-producer franchise the socially optimal arrangement of such research activities. This may be contrasted with the previous case, where research output was assumed to be the sum of the independent, but possibly duplicated, outputs of the research teams, and it was argued that some competition, despite the duplication, might be desirable.

5.6   CONCLUSION

This chapter has argued that technological competition will tend to draw excessive resources into inventive activity. The set of problems thrown up by science and technology represents an open access pool for exploitation by inventors in a process of rent-seeking. But this rent seeking, in fact, results in rent dissipation as each inventor vies to be first by employing more and more inventive resources. The time–cost trade-off, derived in Chapter 2, is the vehicle by which an invention race is modelled in section 5.3, where it is shown that a single inventor believing himself to be threatened by rivals will invent earlier and at a greater cost than if he believes himself not to be under threat. Whether threatened or not, the speed of invention is reduced (along with the cost of invention) if the duration of patent protection is shortened. Since patent protection is an instrument of public policy, this points to the possibility of social control, which is examined in section 5.4, where it is shown that the optimal duration of patent protection is finite. However, if there is also public control over the royalty rate that the inventor can set, it is preferable to give him infinite patent protection but with a royalty which results in later invention than he would choose if he could decide the royalty himself with the public authorities choosing the length of patent protection. Finally, section 5.5 eschews the time–cost trade-off in favour of a simpler fishing paradigm of invention subject to diminishing returns in order to model strategic interaction in the inventive rent-seeking process. Here there is an important distinction to be drawn between a research production function that is specific to the inventor and one that applies to the set of inventors. The latter corresponds to the idea of open-access fishing for research results. Where the research production function is specific to the inventor or research team, Cournot competition between them can lead to increased inventive output compared to that of a single research team using the same resources so long as the teams do not duplicate each other's work. In fact some duplication can be tolerated in exchange for the higher marginal product achieved in smaller teams. Where the research production function applies to the research sector as a whole (so that duplication as such is ruled out), there are increasing research inputs and output as the number of Cournot competitors increases. But this competition erodes away the social gains and, in the limit with free entry, completely. With such a production function it is socially optimal to have just one research team.

NOTES

1. This equivalence was noted by Martin Weitzman (1974):

   A somewhat more fanciful example of a shared resource shows up in some forms of pure research. Considering the stock of potential results in a specific research area as a pool of knowledge which is less than directly proportional to the research effort being applied due to decreasing returns, certain research activities can be cast in the prototype mould [of common property with free access].

   It is also stressed in Brian Wright's (1983) analysis of alternative incentive schemes for invention.
2. This would be valid, for example, if they were always used in fixed proportions.
3. See Chapters 2 and 7 for justification of this assumption of increasing marginal cost in both a priori and empirical terms.
4. The theory of common property resources was initiated by H. Scott Gordon (1954) in the context of an analysis of resource allocation in fisheries.
5. An exception is the early paper of Sir Arnold Plant (1934).

6. This way of posing the social planning problem was first set out by William Nordhaus (1968, Chapter 5). Nordhaus, too, was analysing the optimal period of patent protection, but instead of using a time–cost trade-off he postulated invention production as a function of research inputs. Another important difference between Nordhaus's and the present analysis is that he considered only non-rivalrous invention and ignored the common pool problem.

7. This can be seen as follows: First, the numerator is positive as a consequence of the first and second-order condition for profit maximisation:

$$\partial\pi/\partial T = R_T - C_T = 0 \text{ and } \partial^2\pi/\partial T^2 = R_{TT} - C_{TT} < 0$$

but $R_{TT} = -\rho R_T = -\rho C_T$, hence $\rho C_T + C_{TT} > 0$.

Secondly, the numerator is less than the denominator since $\rho C_T < 0$ and $-(\rho R^0/D).C_T > 0$. Therefore the ratio is bounded by zero and one:

$$0 < \{(C_{TT} + \rho C_T)/(C_{TT} - \rho C_T R^0/D)\} < 1 \rightarrow 0 < \theta < \infty$$

8. See John Panzar and Robert Willig (1977) and William Baumol and Willig (1981), both of which can be found reprinted in Baumol, Panzar and Willig (1982), for the basic articles that establish these propositions.

9. There have been several formulations of technological game-theoretic competition when the time–cost trade-off is stochastic. Examples are Glenn Loury (1977), and Tom Lee and Louis Wilde (1980); these and others are summarised in Kamien and Schwartz (1982, Chapter 5).

10. Thus he states: 'The question which they one and all failed to ask themselves, however, is what these people would otherwise be doing if the patent system were not diverting their attention by the offer of monopolistic profits to the task of inventing (Plant, 1934, p. 40).

11. The acronym coined by Bhagwati stands for 'directly unproductive profit-seeking'. This general area first found application in the field of international trade (Krueger, 1974) but has also been applied to industry studies by Richard Posner (1975). There are obvious applications in the economics of public choice, as the contributions to Buchanan *et al.* (1980) show.

12. Concavity of $P(Z)$ ensures that Z^* locates a maximum.

13. This can be seen by taking the total differential of $n.Z^{**}$, recalling the equilibrium condition 5.9, and setting $dn = 1$. The approximate increase in total research inputs is then:

$$Z^{**} + P'(Z^{**}).(2n - 2)/(P''(Z^{**}).(2n - 1)).$$

where the second term in the sum is negative, thus making the overall sign ambiguous.

14. The elasticity of equilibrium size with respect to numbers is: $(n/Z^{**}).(dZ^{**}/dn) = P'(Z^{**}).(2n - 2)/(Z^{**}.P''(Z^{**}).(2n - 1))$, and comparison with the expression in note 13 will confirm the elasticity proposition in the text.

Part II
Empirical

6 Causality between Inventive and Economic Activity

Chapters 2 to 5 have relied on an assumption that invention is, to a degree, endogenous. The empirical section of this book, beginning with this chapter, aims to examine the question of endogeneity. The relation between the level of economic activity in a sector of the economy and the level of inventive activity directed at that sector is examined in this and the following chapters. This chapter focuses on evidence from time-series of counts of patents, using data originally derived and constructed by Jacob Schmookler. Section 6.1 presents a summary introduction to the history of thought on the question of endogeneity and exogeneity. In section 6.2 the data that Schmookler used, and which is re-examined here, are discussed. Schmookler's assessment of his own data and analysis is examined critically in 6.3. Statistical methods for the analysis of causality in time-series are briefly reviewed in 6.4, and a subset of Schmookler's data is subjected to such methods in 6.5. The summary and conclusions in 6.6 take stock of the question whether inventions cause or are caused by economic variables.

6.1 EXOGENOUS AND ENDOGENOUS VIEWS OF INVENTION

It is normal nowadays to assume that inventive activity is an economic process which absorbs scarce resources and produces, in the aggregate though not necessarily for any particular project, an output that can in principle be given a social valuation. The analyses of the foregoing chapters were based on this assumption. But this view has not always held. At least until a couple of decades ago the dominant view among economists was of invention as a process largely *uninfluenced* by economic forces − exogenous invention. The view that invention is largely exogenous amounts to a contention that the opportunity cost of resources absorbed by inventive activity is virtually zero, either because minimal resources are used or because what are used have no alternative use. One way of rationalising this point of view is to see inventions arising from a kind of spontaneous concatenation of ideas, possibly containing some novel feature from the 'march of science'. This has been called invention by serendipity, or more prosaically the 'supply–push' theory of invention. If this is how inventions arise, and if the nature of the inventive process is such that the conscious pursuit of inventions, by working for them and by allocating resources to their production, is futile, then the only interesting economic questions concern the *development* and *application* of inventions, i.e. innovation.

The serendipity school of thought would argue that the determining forces in the generation of inventions are such things as the growth of science, the spark of genius, the flash of insight, and so on – the kinds of factors that would require sociological, psychological, historical and other non-economics disciplines to explain them. Such an account seems to hinge on a definition of invention that is practically synonymous with 'discovery'. It certainly relegates all connotations of contrivance, design and creation to second-order importance, leaving the principal role to chance. Moreover, it commits the fallacy of composition to argue from the *futility* of seeking particular inventions to the exogeneity of the inventive process in the large.

On the whole, until about 1960, the dominant point of view among economists was that of the serendipity school: that autonomous supply factors were of overriding importance in the generation of technical change. The importance of technical change was not played down but its role was that of a *deus ex machina* – economists were concerned to examine its consequences rather than its causes. Authors examining secular changes in industry who have been associated with this point of view include Joseph Schumpeter (1939),[1] Simon Kuznets (1936), Arthur Burns (1934), Robert Merton (1935) and W. E. G. Salter (1960). Moreover, in the field of macroeconomic growth measurement, the early identification of a large residual element in recorded economic growth which could not be ascribed to increases in conventional factor inputs by Moses Abramovitz (1956) and Robert Solow (1957) tended to perpetuate the view that supply factors were of predominant importance since the residual was for some time identified with technical progress and seemed to imply costless increases in output per unit of input. In growth models technical progress appeared on the supply side as an exogenously given factor.

It may be noted that the exogenous invention point of view is inconsistent with the idea that the creation of a property right in inventions, through a patent system, would stimulate inventive activity. It is also difficult to reconcile it with the fact that a large amount of resources *are* in fact consciously devoted to exploratory research by the private enterprise sector.

The very fact that by the 1960s technological change was assuming a manifest and increasing importance in empirical macro studies, however, was instrumental in focusing attention on the process at the micro level. Economists were clearly concerned about their inability to explain the best part of recorded increases in output over time. What was needed was more factual information and empirical studies.

Important contributions were made, *inter alia*, by Zvi Griliches (1957) and Jacob Schmookler (1962) separately, and together (1963). Griliches showed that the rate of diffusion of innovations could be largely explained by expected profitability considerations. Schmookler, in a major work (1966), attempted to show that the same was true of inventive activity itself. Other authors, of whom the most prolific was Edwin Mansfield (1968), also adopted this point of view with success in empirical analyses of innovations, and further support for the importance of demand factors in successful innovations was provided in case studies by John Jewkes *et al.* (1969) and the OECD (1968).

Of course, expressions of an endogenous invention point of view can be traced to a number of early writers. It is caught by the adage 'necessity is the mother of invention'. René Descartes (1637) reckoned that he could 'find truths in proportion to the efforts [he] made to find them',[2] and Nicolas Rescher (1978) observes that 'it was Leibnitz who first clearly conceived of science as a *productive enterprise*'. Among economists, Adam Smith (1776) recognised that technical improvements would arise from the attempt to increase profits by the division of labour, and Alfred Marshall (1890) maintained that in the long run declining costs can be achieved by a search for new techniques. An early expression of the induced innovation hypothesis is the following quotation[3] from Frank Taussig in 1915: 'the direction in which the contriver turns his bent is immensely affected by the prospect of gain for himself.' The more recent induced invention theories have been discussed in Chapter 4.

In recent times the assumption that inventive activity is

endogenous underlies a sizeable literature that examines what kind of market organisation is most conducive to research and development and invention. This literature is efficiently summarised in Chapter 3 of Morton Kamien and Nancy Schwartz's (1982) survey monograph, in which about 90 separate studies are mentioned, almost all dated post-1960. Prominent among these references is Schmookler's 1966 monograph *Invention and Economic Growth*, which may have been the single most important work to reverse the previously dominant view of invention as exogenous to the economic system.

Schmookler showed that counts of patents – in particular, patents for machinery, equipment and processes – are related to the volume of investment in the industries that use the capital equipment to which the patents relate. He used two kinds of evidence: one based on a sequence of cross-sections of industries, and the other based on long time-series for a small number of particular industries. The cross-section data are examined in Chapter 7. In this chapter the time-series data are used to assess the argument that invention is exogenous: that it affects but is not affected by the level of economic activity.

On the basis of Schmookler's correlations between the level of economic activity and the level of inventive activity, both for the cross-section and the time-series data, it is clear that these variables are strongly associated. The question of exogeneity or causal ordering of the relationship is treated in the next chapter as relating to the elasticity of supply of inventions – an approach that requires the cross-section relationship to be interpreted as identifying a supply curve. The theme of the present chapter is that if there is a temporal sequence in the correlation between these variables, then it may be argued that causality runs from the earlier changing variable to the later changing one.

Schmookler claimed to have identified a lag in the correlation, with the patent data series following the economic activity series, and he argues that 'since causes precede effects, the hypothesis that the observed correlation reflects the effect of invention on investment simply will not do'. (1966, p. 107). Examining this proposition is the theme of the remaining sections of this chapter.

6.2 DATA CONSIDERATIONS

Schmookler assembled long time-series of counts of patents for new technologies applicable in many industries.[4] A subset of these, for which long series measuring the level of investment activity in the corresponding industries could be put together, was used by him to study the relation between economic and inventive activity. He focused attention on three industries: railroads, building and petroleum refining. Of these, the longest time-series showing co-movements between inventive and economic activity could be assembled for the first two, and railroads could be further subdivided into three categories: rails, freight cars and passenger cars. These are examined in detail in sections 6.3 and 6.5, the first of which recapitulates and assesses Schmookler's own analysis, while the latter subjects the same data to more sophisticated statistical testing.

Most of the data used in this chapter are derived from Schmookler (1966), which is the original source of the historical patent statistics, and an account of the compilation of these patent data is given in his Chapter 2. The basic procedure was to develop a concordance between the technological classification used by the US patent office and an industry classification. This concordance was established either from the definition of the sub-classes of technological classification or from direct sampling of the patents if the definition alone left the matter unsettled. For all patents issued after 1874, the counts are on a 'when applied for' basis. For the earlier period, in which the counts are on a 'when granted' basis, the average time-lapse between application and grant was around six months only.

Of course, the patent data are used in this and in following chapters as an indicator of inventions, and since a patent application is an end-product of research or inventive activity, it is clear that the inventive activity took place prior to the patent application. In the context of this chapter, this is an important observation because the concept of causality made use of here is essentially the statistical forecastability of one time-series from another. This leads to an obvious caveat that, even if a patent series can be predicted from some other time-series, it need not necessarily imply that a series measuring inventive activity directly would be so predicted, even if it could be observed.

Another issue is whether a pure count of patents represents an adequate index of invention. Ideally inventions should be weighted by some measure of importance. Of course, not all inventions are patented, nor are all patentable. Of those that are patentable but not patented, the vast bulk are probably inventions of very minor importance. Establishing a property right in an idea through a patent is not a costless process, so it may be assumed that patented inventions are on average more important than non-patented but patentable inventions. Hence an element of crude weighting, one and zero, is implicit in the use of patents as a proxy for patented inventions. None the less, as will be emphasised in Chapter 8, the distribution of patented inventions by value is very spread out. In such circumstances the number of patented inventions may not reliably measure their value, so patent counts may be nearer to a measure of inventive inputs than to a measure of inventive output. This seems to be borne out by the stability of patent/research ratios over time in particular industries.

6.3 SCHMOOKLER'S ASSESSMENT OF THE TIME-SERIES

This section examines in detail the relation between counts of patents for new capital goods in United States railroads and the level of investment in that industry. In addition to the aggregate, three types of new investment, and patents relating to them, are examined. They are: rails, freight cars and passenger cars, respectively.

The data are presented in Figures 6.1 to 6.4. From examination of these it is clear that there is co-movement between the invention index series and the series on capital formation. For the aggregate series Schmookler separated the trend from what he calls the 'long swings' or Kuznets cycles. The trend was established by a 17-year moving average and the long swings were exhibited as 7- or 9-year moving averages of the year-by-year deviations from the trend.

As far as the 'trend' is concerned, economic activity and inventive activity display an inverted U-shape, and both trend series reach a peak around 1910. The long-swing picture is, if anything, even more impressive as an exhibit of co-movement between the series. The data on fixed investment only start at 1870, so to extend

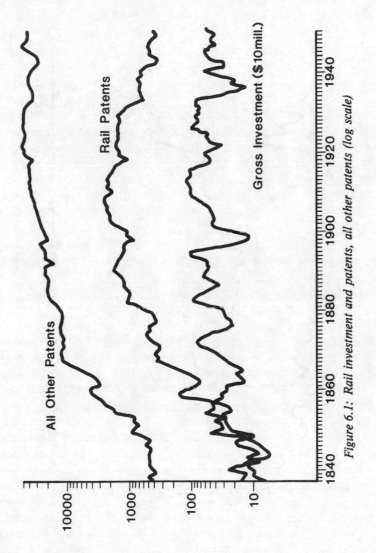

Figure 6.1: Rail investment and patents, all other patents (log scale)

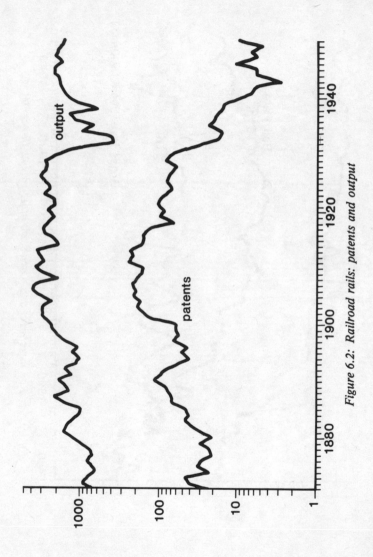

Figure 6.2: Railroad rails: patents and output

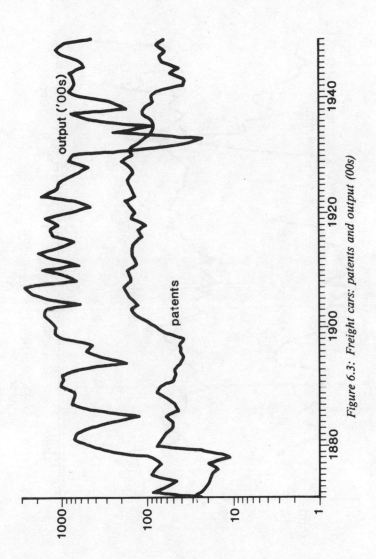

Figure 6.3: Freight cars: patents and output (00s)

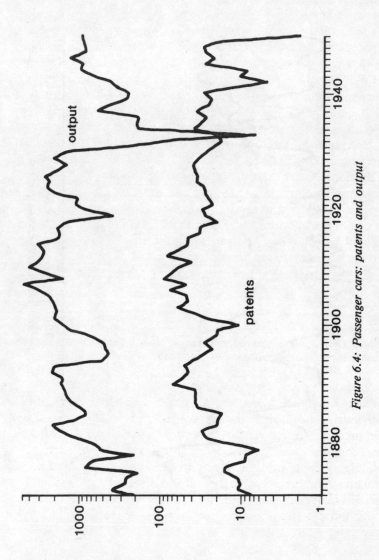

Figure 6.4: Passenger cars: patents and output

it further back in time Schmookler uses an index of the number of miles added to the railroad network.[5] Figure 6.5 shows clearly the effectiveness of the smoothing decomposition as a device to produce harmonic oscillations between the series.

The co-movements may, however, depend on the smoothing procedure used. To test this, a small experiment was conducted with a robust smoother, using running medians[6] as proposed by J. W. Tukey (1977). This is displayed in Figure 6.6. A chronology of turning-points from these smoothed series is slightly different from that obtained by Schmookler's method. However the differences would have little effect on Schmookler's argument.

Schmookler was, however, not content merely to have demonstrated that these series have similar trend and long-swing behaviour, but he argued that patents usually lag behind the economic indicators at the main turning-points.

This question of phasing is, of course, critically important when it comes to establishing which of these variables might cause or be caused by the other. As Nathan Rosenberg (1974), in his review article of Schmookler's book, states:

In examining the railroad industry, for which comprehensive data are available for over a century, Schmookler found a close correspondence between increases in the purchase of railroad equipment, and slightly lagged increases in inventive activity as measured by new patents on such items. The lag is highly significant because, Schmookler argues, it indicates that it is variations in the sale of equipment which induce the variations in inventive effort.

When Rosenberg describes the lag as 'highly significant' he obviously means this to apply to the role of the lag in establishing Schmookler's argument that inventions are endogenous – induced by economic variables. It is not at all clear, however, that the lag is 'highly significant' in the statistical sense. The question of statistical significance is now examined briefly, first for all railroad patents and investment, and then for the three sub-fields of railroad rails, passenger cars and freight cars, respectively.

(i) All railroad patents and investment
If the patent series is compared with the investment series (new additions to road spliced on the gross fixed capital formation data),

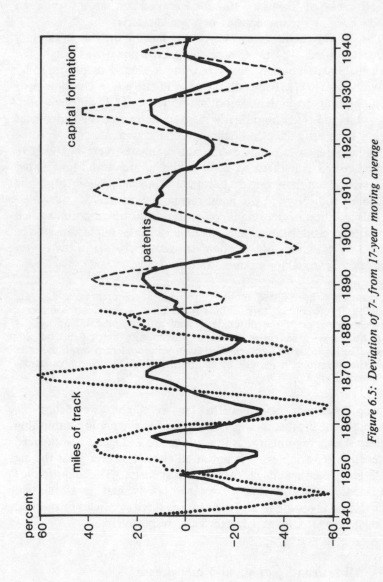

Figure 6.5: Deviation of 7- from 17-year moving average

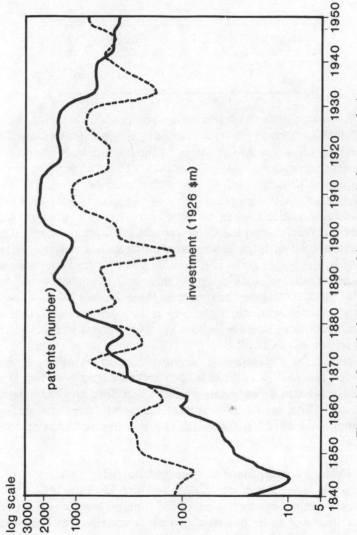

Figure 6.6: Smoothed patents and investment (robust smoother)

then the following table can be constructed:

Table 6.1

	Peak	Trough	Total
Patents lead	1	2	3
Investment leads	4	2	6
The two tie	0	2	2
Total	5	6	11

These data invite the following question: with what probability would this pattern be observed if leads by either series were equally likely? Ignoring ties and referring to the binomial distribution, it appears that at least six out of nine occurs about 25 per cent of the time and at least four out of five with probability around 0.19.

But Schmookler uses his skills as an historian to cast doubt on the three occasions when patents lead – during the Spanish-American War, during the Civil War and the third associated with a dubious lead in 1905. Schmookler's approach is only to believe a lead by the patent series if it is confirmed by a lead over a series relating to railroad stock prices. In this way patents only lead once out of eight turning-points (omitting the dubious 1905 case), and that would be statistically significant at the 5 per cent level, though Schmookler does not use the concept of statistical significance in this section of his book.

Apart from appearing to be introduced in a rather *ad hoc* manner, the stock price series is open to other objections. This kind of series is based on expectations and is therefore forward-looking in nature. The use of any expectations-based data must bias an attempt to identify leads and lags as a way to establish a case for causality.

(ii) Railroad equipment output and patents

For these series Schmookler does not use his ratio of moving averages method to identify peaks and troughs, but picks them out from the raw data to give what he calls 'a tentative but somewhat arbitrary list' of peaks and troughs. The information is summarised in Table 6.2.

Ignoring ties, the probability of as many as 23 output leads out of 33 instances of one or other series leading at a turning-point is

Table 6.2

	Rails		Passenger Cars		Freight cars		Total
	Peak	Trough	Peak	Trough	Peak	Trough	
Patents lead	2	0	3	2	2	1	10
Output leads	3	5	4	4	2	5	23
The two tie	1	1	1	1	2	0	6
Total	6	6	8	7	6	6	39

less than 2 per cent assuming that leads and lags are equally likely. Notice, however, the significantly different pattern between peaks and troughs. Output is very clearly the leader at the lower turning-point in most instances; but there is not much in it at the upper turning-point (patents lead 7 times and output leads 9 times).

Thus a clear pattern seems to emerge for these sub-series, and it reverses the slight tendency in the all railroad patent/railroad investment series for the phase difference to be concentrated at the peaks. Here it is very definitely in the troughs that the precedence of economic activity shows through.

Schmookler did not even attempt to apply any statistical tests of inference to his data. Perhaps he was only too conscious of the tentativeness and arbitrariness of his listing of peaks and troughs. He was content simply to have demonstrated that the data are suggestive of a causality pattern running from economic variables to inventive activity. But, it might be asked, how tentative and how arbitrary is the list of peaks and troughs? And how much of a Slutsky effect[7] is there in Figure 6.5 showing synchronous oscillations?

Beyond these questions there is a more fundamental objection to Schmookler's methodology, namely that turning-point analyses are very inefficient ways of using time-series data which oscillate at low frequencies. The long swings in these data are of about 15 years' duration, so only about 10–15 per cent of the information in the data series is used in turning-point analysis.

Section 6.5 accordingly re-examines Schmookler's data in ways that avoid both smoothing and arbitrary dating of turning-points while at the same time using all the information in the time series. Before getting there it is necessary to introduce the relatively recent time series concept of 'Granger causality'.

6.4   THE TIME-SERIES APPROACH TO CAUSALITY

The idea that causes precede effects underlies a modern statistical procedure for examining 'causality' relations between variables. The very terminology of cause and effect can quickly lead to deep philosophical issues of course, but suffice it to note here that even in physical systems where the language of cause and effect gives rise to fewest difficulties, it is not universally true that observable causes precede observable effects. The first sign of a causal fire may well be its effect of smoke because there are differential observation lags for the two phenomena. In social systems, and especially in economic affairs where successful anticipation can be highly rewarded, this is compounded by the fact that expectations influence behaviour and hence that observations about the present are partially determined by current views about the future. Since current opinions are usually unobserved, it follows that the appearance may be given that future observables, which in truth are exogenous, are caused by the current set of observed variables that are affected by expectations about the future.[8] This caveat about interpreting time sequences applies as much to the statistical procedures to be outlined in this section as to Schmookler's simpler approach reported in the previous section. If, however, it can be assumed that future observables cannot cause the past observations, then the predictability of future values of one time-series from past values of another series would seem to be a necessary if not sufficient reason to argue that the latter series causes the former.[9] This is causality in the sense of C. W. J. Granger (1969).

Variable x is said to cause variable y in the sense of Granger if knowledge of past values of x can improve the prediction of y conditional on all past information in the universe other than on x. Prediction improvement is judged by a smaller mean square error of the forecast. To put this definition symbolically, let:

   y^* be the predicted value of y,
   X' be all past information on x,
   X be all past and present information on x, .
   U' be all past information in the universe other than on x,

and U be all past and present information in the universe other than on x.
Then Granger defines:

Causality:
If $\text{var}(y^*|U',X') < \text{var}(y|U)$
then x is said to cause y

Instantaneous causality:
If $\text{var}(y^*|U,X) < \text{var}(y|U)$
then x is said to cause y instantaneously.

As it stands this definition of causality is non-operational since U or U' is an unattainable ideal. Granger however suggests replacing 'all information in the universe' by the more manageable definition: 'all relevant information'. Obviously, judgement based on some kind of *a priori* notion must be used to decide what is relevant.

The tests of causality that have been proposed in the literature fall into two broad categories, namely, tests based on a suitable cross-correlation function and regression-based test procedures. For those in the former category the method is to examine the significance of leading and lagging cross-correlations[10] between the pre-whitened x and y series. Such methods have been proposed by David Pierce (1977) and Larry Haugh and George Box (1977). There are however ambiguities in this approach concerning the pre-whitening filters. Should the same filter apply to both the x and y series, or should the series be filtered separately? If different filters are applied it is likely that some of the causality relation will be filtered away, biasing such tests against detection of Granger causality.[11] On the other hand, assuming that the spectral distribution of both x and y cannot be flattened by the same filter, there is a problem in choosing the common filter. An *ad hoc* approach adopted by Christopher Sims (1972) was to use Nerlove's 'universal filter',[12] but subsequent researchers have preferred to use filters specific to the given time-series. Perhaps the cross-correlations should be calculated for data filtered by the apparent ARMA structure of both series, thus producing two tests. A further difficulty with the cross-correlation approach is that it allows no place for specific information in the set U that is not contained in past values of x or y. In other words it implicitly assumes that all relevant information is contained in the histories of x and y.

Regression procedures are based on an assumption that the time-series processes for x and y can be represented by a jointly autoregressive scheme:

$$
\begin{bmatrix} a_{11}(L)\, a_{12}(L) \\ \\ a_{21}(L)\, a_{22}(L) \end{bmatrix}
\begin{bmatrix} y_t \\ \\ x_t \end{bmatrix}
=
\begin{bmatrix} b_{11}(L) \ldots b_{1k}(L) \\ \vdots \\ b_{21}(L) \ldots b_{2k}(L) \end{bmatrix}
\begin{bmatrix} z_{1t} \\ \vdots \\ z_{kt} \end{bmatrix}
+
\begin{bmatrix} e_{1t} \\ \\ e_{2t} \end{bmatrix}
\qquad (6.1)
$$

in which $a_{ij}(L)$ and $b_{ij}(L)$ represent lag polynomials, the z_i are exogenous variables and the e_i are disturbance terms. If it could be established for example that $a_{12}(L) \neq 0$ while $a_{21}(L) = 0$, this would imply that x causes y and is not itself caused by y, thus establishing unidirectional Granger causality from x to y. To establish causality patterns by regression methods the analysis might be essentially non-parametric or it could attempt to model the joint autoregressive relationship by parametric methods. One advantage of parametric procedures is that they yield parameter estimates: the coefficients of the lag polynomials and their implicit impact and long-run multipliers or elasticities, which could be of independent interest.[13] However, such methods are computationally more demanding and are likely to be less robust, due to possible specification error, than the essentially non-parametric regression methods, two of which are detailed below.

The ability of regression-based procedures to incorporate exogenous variables is a particular advantage over the cross-correlation approach. They allow for joint causation, admitting influences on y other than through x and which are not included in y's own past history. However y's past history should also incorporate influences due to past changes in the exogenous variables, so in truth only current values of the z_i variable matter from a practical point of view.

The regression-based tests of Granger causality used in the following section are both detailed in Harvey (1981, Chapter 8), which presents a 'direct test' and an 'indirect test'. For either test the null hypothesis that x does not cause y will be rejected if it can be shown that there is information in the history of the x series which is special to the x series and which allows an improved prediction of y. Both tests assume that the bivariate time series (x_t, y_t) is linear, covariance stationary and purely non-deterministic.

The *direct test* can be explained by writing out the equations 6.1 with only x_t and y_t on the left side, assuming that only current values of the exogenous variables matter:

$$y_t = \sum_{j=1}^{n} a_{11j} y_{t-j} + \sum_{j=0}^{m} a_{12j} x_{t-j} + \sum_{j=1}^{k} b_{1j} z_{jt} + e_{1t}$$

$$x_t = \sum_{j=0}^{n'} a_{22j} x_{t-j} + \sum_{j=1}^{m'} a_{21j} y_{t-j} + \sum_{j=1}^{k} b_{2j} z_{jt} + e_{2t} \qquad 6.2$$

To examine the null hypothesis that x does not cause y, the first equation of 6.2 suggests regressing y_t on $y_{t-1}, y_{t-2}, \ldots, y_{t-n}, x_t, x_{t-1}, \ldots, x_{t-m}, z_{1t}, \ldots z_{kt}$ and testing the joint significance of the m lagged values of x_t.[14] The n lagged values of y_t, the k exogenous regressors z_{jt} and the disturbance term e_{1t} should contain all influences on y_t other than the putatively causal x series. If there are systematic influence on y_t not allowed for in the equation, these would show up in the disturbance term which would therefore presumably depart from white noise. Harvey (1981) therefore emphasises that the value of n, the longest lag on the lagged dependent variable, should be chosen sufficiently large that such stochastic autoregressive features be eliminated. In practice the direct test is carried out by calculating two regression equations, including and excluding the lagged x values, and performing a subset F-test to test the significance of the m lagged x values.[15]

The rationale for *indirect tests* is due to Sims (1972):

We can always estimate a regression of Y on current and past X. But only in the special case where causality runs from X to Y can we expect that no future values of X would enter the regression if we allowed them. Hence we have a practical statistical test for unidirectional causality: regress Y on past and future values of X, taking account by generalised least squares or prefiltering of the serial correlation in the error term. Then if causality runs from X to Y only, future values of X in the regression should have coefficients insignificantly different from zero, as a group.

The regression model is therefore:

$$y_t = \sum_{i=n}^{-m} a_i x_{t-i} + e_t \qquad 6.3$$

and Sim's proposal is to regress y_t on past, present and future values of x_t, testing the joint significance of the future x_t variables. An important assumption is that the error term e_t is a white noise process – if this assumption is violated then the results of the subset F test might be misleading. This was recognised by Sims in his own application in which he pre-filtered both series by the filter referred to in footnote 12. This *ad hoc* filtering is avoided in a procedure

due to J. Geweke *et al* (1982).[16] Suppose that e_t is serially correlated
of order p, then including p lagged regressors on the right hand side
of regression equation 6.3 should effectively wipe out the serial
correlation, leaving the subset F test valid. Hence the causality tests
using Sim's indirect method reported in the next section assume
that any time series structure in the residuals e_t of equation 6.3 can
be adequately approximated by an autoregressive process of order
p. They do so by including p lagged regressors and p further lags
on the x variable in addition to the n represented in equation 6.3.

6.5 RESULTS OF CAUSALITY TESTS

The direct and indirect regression methods of testing for Granger
causality outlined in the previous section are applied to
Schmookler's time-series data for the United States railroad and
building industries. The former industry is examined in greater
detail, partly because a longer time-series is available, but also
because Schmookler has provided data for three sub-categories of
capital formation in addition to the aggregate data for that
industry.

(i) Railroad patents and railroad capital formation
All variables except time are defined as differenced logarithms
(i.e. rates of change), and the concomitant (exogenous) variables
are W, real fixed investment in manufacturing industry,[17] and V,
all patents other than railroad patents. The 'causal' and 'caused'
variables are interchangeably patents and real investment in
railroads. The investment data, which Schmookler presents for
1870 to 1950, were extended backwards to 1837 by splicing them on
to Schmookler's data for net additions to miles of railroad using
a simple linear regression for the overlapping years 1870 to 1892.
    Examination of the log-differenced patent data revealed a hetero-
skedastic series with the variability greater in the period prior
to the 1860s than thereafter.[18] This can be seen in Figure 6.7.
Accordingly, the regressions were calculated not only for the full
period of data availability, but also excluding the first ten observations. The spectral density of the log-differenced patent series is
displayed in Figure 6.8, from which it is evident that there is
time-series structure at low, middle and high frequencies. The

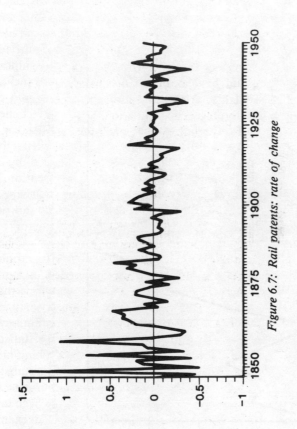

Figure 6.7: Rail patents: rate of change

dominant lowest frequency density indicates that differencing did not eliminate long-run 'trend' elements. Since the causality relation between these series, if any exists, is not expected to be of such a long-range nature, the trend was allowed for in the regressions by

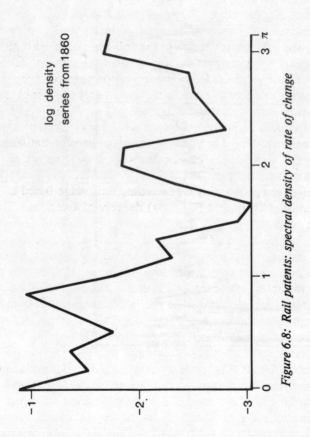

Figure 6.8: Rail patents: spectral density of rate of change

including time as a variable. The middle-frequency peaks in the spectral density indicate periodicities of 8 and 3 years, respectively. The former appears to correspond with the long swings in investment, and the latter with a peak at a periodicity of about $3\frac{1}{2}$ years in the investment series.

The regressions are summarised in tables 6.3 to 6.5. Table 6.3 presents the essential information for assessing Granger causality on the basis of the *direct test*. Each row of the table presents information from two regression equations: one with and the other without nine lagged values of the 'causal' variable. Each section of the table presents all six possible combinations of three concomitant variables, showing their t ratios in the full regression with all lags of the 'causal' variable present. In all cases the current unlagged 'causal' variable was included in the regression equation. Hence the regressions seek to examine jointly the hypotheses of strict causality and instantaneous causality. All equations were estimated with both the full data series and the data series excluding the first ten effective observations. The final two columns present statistics for assessing the significance of the lagged 'causal' variable. The penultimate column gives the subset F statistic while the final column gives the probability of exceeding that value based on the F distribution with m and (T-n-m-k) degrees of freedom.

On the basis of the full data series the results are somewhat equivocal in the sense that the F statistics look broadly similar for causality in both directions, with the level of significance hovering in the 10 per cent region, depending on the set of included concomitant variables. Instantaneous causality, assessed by the t ratio of the current causal variable, is not established by these regressions except when the two trend variables (constant term and time) are included in the regressions. This version of the equation is also the most favourable in establishing strict causality in either direction because the null hypothesis that the lagged causal variables do not improve the fit of the equation is rejected at the 5 per cent level.

The assessment of causality is markedly different when the first ten observations are excluded. Now the lagged causal variables achieve significance at the 5 per cent level (in the best case at the 1 per cent level) for causality running from investment to invention; and absence of causality in the opposite direction cannot be rejected (in the best case the lagged causal variables just achieve significance at the 10 per cent level). Evidence for instantaneous causality is strengthened compared with that for the full data set. It appears however that this instantaneous causality is affected by the amount of investment carried out in manufacturing industry since inclusion of this variable in the regressions with railroad investment as the dependent variable results in a substantially lowered t statistic for the current number of railroad patents. An explanation for this might be that an application for a patent is itself a form of investment, and hence there may be common factors affecting these variables such as the general climate of business expectations, interest rates and the availability of finance. To the extent that capital formation in manufacturing captures these common elements it reduces the contemporaneous partial correlation between investment and patenting in railroads. In summary, in direct test for causality between railroad patents and railroad investments suggests one-way causality from investment to patents with a hint of instantaneous causality.

Tables 6.4 and 6.5 present information to assess Granger causality between railroad patents and railroad investment on the basis of the *indirect test* using Sim's (1977) theorem. The layout of Table 6.4 corresponds to that of 6.3, being divided vertically into two main parts according to the direction in which causality is being tested.

With these indirect tests it must be borne in mind that the role of the dependent variable differs from its role in the direct test – here it represents the putatively 'causal' variable. Thus the top section of both tables explores the possibility that causality runs from investment to patents on the basis of the full data set.

The implications of Tables 6.4 and 6.5 are much less ambiguous than those of the preceding table relating to the direct test. Again, the preferred dataset is that excluding the first ten observations, and for those data the indirect test clearly rejects the null hypothesis that future patents do not improve the prediction of current investment, but cannot reject the hypothesis that future investment does not improve the regression prediction of current patents. In other words, causality is established from investment to patents but rejected in the reverse direction. In the former case there is also the suggestion of instantaneous causality for the best fitting equations, where the t statistic on current patents just exceeds the 95 per cent confidence level.

The precise form of the indirect test is a compromise between

Table 6.3: Railroad patents and railroad capital formation: the direct test

Dependent ('caused') variable	Data	Concomitant variables				Causal variable		P(x>F)
		Const. (t ratio)	Time (t ratio)	V (t ratio)	W (t ratio)	Current (t ratio)	Lagged values (F ratio)	
Patents	1852–1950	0.07				1.49	1.97	0.054
"	"				6.73	1.47	1.94	0.058
"	"	−1.42			6.91	1.29	1.51	0.160
"	"	3.68	−3.72			1.33	1.55	0.146
"	"	1.22	−1.44		5.60	1.97	2.59	0.023
"	"					1.54	1.58	0.136
"	1862–1950	−0.18				1.95	2.29	0.026
"	"				7.00	1.95	2.26	0.028
"	"	−1.75			7.32	2.49	2.00	0.053
"	"	3.05	−3.10			2.61	2.08	0.043
"	"	1.66	−1.89		6.59	2.68	2.56	0.013
"	"					3.01	2.16	0.036
Investment	1852–1950	0.29				1.43	1.92	0.060
"	"			2.79		1.47	1.87	0.068
"	"	−0.12		2.76		1.03	1.69	0.105
"	"	−1.52	1.59			1.02	1.67	0.109
"	"	−1.38	1.38	2.62		1.97	2.08	0.041
"	"					1.47	1.83	0.075
"	1862–1950	0.40				1.95	1.11	0.367
"	"			2.94		1.95	1.08	0.387
"	"	−0.07		2.88		1.35	1.09	0.384
"	"	−2.50	2.57			1.34	1.07	0.394
"	"	−2.28	2.29	2.63		2.68	1.73	0.099
"	"					2.02	1.59	0.136

Note: All variables are expressed as differenced logarithms. V is manufacturing investment; W is all patents other than railroad patents. Lag Pattern: m = 9, n = 9.

Table 6.4: Railroad patents and railroad capital formation: indirect test

Dependent ('causal') variable	Data	Concomitant variables				Caused variable		P(x>F)
		Const. (t ratio)	Time (t ratio)	V (t ratio)	W (t ratio)	Current (t ratio)	Lagged values (F ratio)	
Investment	1852–1950	−0.27				3.33	2.19	0.037
	"				2.34	3.32	2.17	0.039
	"	−0.47			2.36	2.64	1.69	0.114
	"	−1.16	1.13			2.66	1.68	0.118
	"	−1.14	1.08			3.26	2.23	0.034
	"				2.33	2.75	1.77	0.096
	1862–1950	−0.52				1.20	3.26	0.003
	"				2.37	1.25	3.25	0.003
	"	−0.77			2.42	0.56	2.46	0.021
	"	−1.91	1.86			0.63	2.46	0.021
	"	−2.19	2.12			2.21	3.37	0.003
	"				2.63	1.91	2.82	0.009
Patents	1852–1950	1.40				3.17	1.60	0.138
	"			8.31		3.02	1.54	0.157
	"	−0.30		8.05		2.70	1.50	0.172
	"	4.73	−4.47			2.69	1.49	0.176
	"	2.22	−2.42			2.49	1.68	0.117
	"			6.61		2.45	1.49	0.174
	1862–1950	1.02				0.90	0.91	0.512
	"			7.82		0.87	0.91	0.516
	"	−0.61		7.68		0.43	0.84	0.573
	"	4.83	−4.68			0.43	0.84	0.573
	"	2.03	−2.23			0.98	1.08	0.390
	"			5.88		0.56	0.81	0.599

Note: All variables are expressed as differenced logarithms. W is manufacturing investment; V is all patents other than railroad patents. Lag pattern: m = 8, n = 5, p = 1.

Table 6.5: Railroad patents and railroad capital formation: indirect test

Dependent ('causal') variable	Data	Concomitant variables				Caused variable		
		Const. (t ratio)	Time (t ratio)	V (t ratio)	W (t ratio)	Current (t ratio)	Lagged values (F ratio)	P(x>F)
Investment	1862–1950	0.03				1.43	4.14	0.001
"	"				2.76	1.42	4.08	0.002
"	"	−0.30			2.76	0.81	3.32	0.006
"	"	−1.79	1.80			0.82	3.26	0.007
"	"	−2.09	2.07		2.94	2.22	4.27	0.001
"	"					1.87	3.47	0.005
Patents	1862–1950	1.13				1.01	0.98	0.445
"	"			7.26		0.97	0.92	0.485
"	"	−0.33		7.07		0.73	0.88	0.514
"	"	5.53	−5.36			0.73	0.87	0.520
"	"	3.22	−3.39	5.35		1.09	1.10	0.373
"	"					0.88	1.00	0.430

Note: All variables are expressed as differenced logarithms. W is manufacturing investment; V is all patents other than railroad patents.

allowing the largest possible number of leads for the caused variable (m in the tables), and eliminating sufficient autoregressive lags in the residual with a suitably high p value. The regressions reported in Table 6.4 allow eight leads but only a first-order autoregressive residual. By contrast the specification of the regression equations underlying Table 6.5 allows six leads but a third-order autoregressive process in the residual. However the qualitative conclusions for this table remain the same: causality is established from investment to patents but not from patents to investment. In fact, the probability may be as low as one in a thousand that the statistic implying causality in these time series could have occurred by chance; in any event the future patent variables are significant at the 1 per cent level in all specifications of the regression equation.

The unambiguous conclusion from the foregoing that there is unidirectional causality from investment to patents in US railroading is explored further in Table 6.6. Two questions are examined in that table about the nature of 'causality' from investment to patents. First, there is the question whether the relation holds for different types of new technology. To examine this railroad patents are divided between track patents and non-track patents. The other question relates more deeply to the underlying idea that it is inventions, as opposed to patents that are caused by variations in the extent of their possible application.

A patent is of course an end-product of inventive activity. The inventive R & D activity that produced the invention must have preceded the data at which a patent was applied for. It can therefore be argued that the underlying inventive activity series leads the patent series and hence that causality tests based on the latter are misleading. Such tests might, for example, incorrectly designate instantaneous causality as strict causality or even apparently detect causality in the wrong direction. To explore this possibility, Table 6.6 includes variants of the indirect test in which the patent series is effectively lagged by one year as well as the usual unlagged tests.

The significance tests reported in Table 6.6 imply first that the overall causality relation between investment and patents reflects an even stronger connection between investment and patents for non-track capital goods combined with an absence of causality in this direction for track patents. This surprisingly clear dichotomy is

Table 6.6: Railroad patents and capital formation

Indirect tests for causality from investment to invention: — allowing for 1-year delay between invention and patent; — railroad patents split between track and non-track patents. Significance tests: t or F ratios and $p(x > F)$

Regressors	All patents		Track patents		Non-track patents	
U, t ratio						
V, t ratio	2.84		3.55		3.32	
No delay (m, n, p)	(6, 5, 3)	(6, 5, 3)	(6, 3, 2)	(6, 3, 2)	(6, 3, 2)	(6, 3, 2)
— Current patents, t	2.50	3.25	0.25	0.91	2.99	3.40
— 6 Future patents, F	1.97	2.26	0.94	0.93	3.31	3.38
— 6 Future, $p(x > F)$	0.080	0.046	0.478	0.475	0.006	0.005
Delay (m, n, p)	(5, 5, 4)	(5, 5, 4)	(5, 3, 3)	(5, 3, 3)	(5, 3, 3)	(5, 3, 3)
— Current patents, t	0.68	1.15	−0.23	0.13	0.91	1.28
— 5 Future patents, F	2.35	2.60	1.12	1.11	3.96	3.99
— 5 Future $p(x < F)$	0.048	0.031	0.356	0.359	0.003	0.003

Notes: The significance tests reported in each column were calculated from a regression model:

$$y_t = \sum_{i=1}^{n} a_i y_{t-i} + \sum_{i=0}^{n+p} b_i x_{t-i} + \sum_{i=1}^{m} c_i x_{t+i}$$

In the 'no delay' version the t test is on b_0 and the F test is on the m c_i coefficients as a group. In the 'delay' version the t test is on c_1 and the F test is on the m − 1 remaining c_i coefficients as a group.

Concomitant variables are U: investment in manufacturing; V: investment in equipment in manufacturing.

All variables are measured as differenced logarithms.
Data: 1850 to 1950.

investigated further in the next sub-section. The other feature examined in Table 6.6, viz. the implications of delaying the patent series by one year to represent better the inventive activity that underlies those patents, is revealed by comparing the probabilities of exceeding the calculated F ratio under the null hypothesis of absence of significance of the future variables for both specifications of a given equation. In every case that probability is lower under the delay specification. This strengthens the conclusion that inventions are caused by investment.

(ii) Railroad rails patents and output

The implication of the foregoing, that the clear unidirectional causality pattern for railroad patents is entirely due to non-track patents and investment and is actually weakened by the inclusion of track patents and investment, is followed up now by examining in more detail an important category of 'track', namely railroad rails. For this sub-category Schmookler's data provide a more specific investment series in the form of output in tons of rails. It is possible that the time-series of overall railroad investment is not a good indicator for this particular category of investment. The statistical price paid for this extra and more specific detail is a shorter time-series by some 35 years.

Despite the obvious co-movements between railroad rails patents and output evident in Figure 6.2, it was not possible to establish unambiguously a causal relation between these time-series. Table 6.7 reports the regression results for the indirect test using the full span of data. There is clearly a suggestion of instantaneous causality but the insignificance of future variables implies that the null hypothesis of absence of strict causality in either direction cannot be rejected.

Since both data series are quite unevenly volatile, there is a possibility that these regression results are strongly influenced by a few extreme fluctuations. As an informal experiment to examine this possibility, the regressions were recomputed from a correlation matrix that was formed from only those observations which did not exceed 0.6 in absolute value for the logarithmic changes. More extreme fluctuations were treated as missing values for the pairwise zero-order correlations, the aim being that the regressions based on such a correlation matrix should represent the main body of the data showing 'normal' fluctuations.[19] Results of these alternative

Table 6.7: Railroad rails patents and rails output: indirect test

Dependent ('causal') variable	Data	Concomitant variables				Caused variable		
		Const. (t ratio)	Time (t ratio)	V (t ratio)	W (t ratio)	Current (t ratio)	Future values (F ratio)	P(x>F)
Output	1875–1950	1.29				2.27	0.84	0.547
"	"				6.45	2.45	0.98	0.449
"	"	1.02			6.31	2.36	1.42	0.223
"	"	0.54	−0.21			2.49	1.52	0.190
"	"	1.18	−0.94			2.29	0.78	0.590
"	"				6.37	2.14	1.21	0.314
Patents	"	−0.96				3.11	1.32	0.284
"	"			4.37		3.17	1.23	0.307
"	"	−1.08		4.37		2.08	1.65	0.150
"	"	1.94	−2.31			2.16	1.56	0.178
"	"	0.94	−1.28			2.53	1.71	0.137
"	"			3.78		1.90	1.75	0.128

Note: All variables are expressed as differenced logarithms. W is railroad capital formation; V is all railroad patents other than rails.
Lag pattern: $m = 6$, $n = 5$, $p = 3$.

Table 6.8: Railroad rails patents and rails output: modified indirect test

Dependent ('causal') variable	Concomitant variables					Caused variable		
	Const. (t ratio)	U (t ratio)	V (t ratio)	W (t ratio)	X (t ratio)	Current (t ratio)	Future values (F ratio)	P(x>F)
Output	2.20					1.57	1.04	0.396
"		3.87				2.13	1.69	0.164
"	1.80					0.94	1.66	0.172
"			5.80			2.32	3.63	0.010
"		2.59	5.05			1.40	3.05	0.024
Patents	−2.11					1.23	0.25	0.907
"	−2.41					1.62	0.13	0.971
"	−2.11				3.67	2.05	0.25	0.909
"				4.34		1.03	0.35	0.844

Notes: (1) The regressions reported above were computed from a correlation matrix formed from only those observations that did not exceed 0.6 in absolute value for the differenced logarithms. This corresponds to a range of +82 per cent to −55 per cent for changes in the untransformed data. The 'test statistics' are therefore not strictly valid, and should be considered merely indicative.

(2) Only regressions in which the concomitant variables are 'significant' are reported.

All variables are expressed as differenced logarithms. U is investment in manufacturing; V is railroad capital formation; W is all railroad patents other than rails; X is total patents other than railroad patents. Lag pattern: m = 4, n = 4, p = 2.

Data: 1875–1950.

regressions are presented in Table 6.8, from which it can be seen that eliminating the influence of extreme fluctuations has the effect of bifurcating the strict causality implications, suggesting once more that strict causality runs from output to patents, and not vice versa. The implications of instantaneous causality are at the same time weaker than previously.

(iii) Rolling stock patents and outputs

Despite the unambiguous causality relationship established in section 6.5(i) between non-track patents and railroad investment, it did not prove possible to trace this down to the more detailed time-series for rolling stock displayed in Figures 6.3 and 6.4. A number of regression experiments were conducted for freight and passenger cars output and patents, but the apparent co-movements exhibited in the diagrams did not carry over to significance in the causality tests between these pairs of time-series. A representative table, presenting the indirect test results for freight car patents and output is presented in Table 6.9. It is clear that the null hypothesis that future variables have no explanatory value cannot be rejected for any of the regression specifications presented in that table. Moreover, the null hypothesis of absence of instantaneous causality cannot be rejected.

(iv) Building patents and building activity

Causality test regressions were also carried out for the only non-railroad time-series that Schmookler presented that was sufficiently long and homogeneous to justify such regressions, namely for building patents and activity. Results of these regressions are not presented here, except to remark that in no specification that was tried could any causality pattern be established.

Both direct and indirect tests were performed with various subsets of the data and a number of regression specifications. In his book, Schmookler suggests that the share of building activity in total capital formation is related to the share of building patents in total patents, but this too yielded negative results.

6.6   DISCUSSION AND CONCLUSIONS

The results of the causality tests reported in section 6.5 are mixed. To the extent that a causal ordering is established it is from investment to patents, it is unidirectional, there being no evidence of

Table 6.9: *Freight car patents and freight car output: indirect test*

Dependent ('causal') variable	t (t ratio)	t^2 (t ratio)	Concomitant variables U (t ratio)	V (t ratio)	W (t ratio)	Caused variable Current (t ratio)	Future values (F ratio)	P(x>F)
Output			3.86			0.06	1.00	0.425
"	−2.06		4.40			−0.08	1.42	0.233
"	−2.41	1.89	4.76			0.00	0.92	0.478
"	3.33	−3.01				0.02	1.50	0.209
"				4.31		−0.34	0.37	0.865
"				2.75		−0.07	0.85	0.521
"			4.96	4.03		−0.33	1.27	0.292
Patents	2.34				2.86	0.30	0.58	0.717
"						−0.17	0.49	0.786

Notes: (i) The variables are defined as (log) levels, so a quadratic trend was allowed for. U is investment in manufacturing; V is investment in railroads; W is railroad patents; t is a time trend. Lag pattern: m = 5, n = 5, p = 2.

(ii) Only those regressions for which concomitant variables are 'significant' are reported.

Data: 1884–1950

feedback from patents to investment, and it is strengthened if patents are lagged by one year more nearly to represent the inventive activity that gave rise to them. These results apply to railroad patents taken as a whole, and more particularly to railroad non-track patents. They do not apply to railroad track patents nor to the relationship between building patents and building activity.

It may be that there is a different relationship between invention and investment for long-lasting investments like structures and earthworks from that for equipment and machinery. Perhaps for long-lasting investments, capital goods invention is determined principally on the 'supply side' and appears exogenous rather than induced. Whether or why this might be so has not been explored in this chapter, but suffice it to say that these negative causality results imply that it is probably not universally true that investment causes invention, despite the very substantial support that the hypothesis gets from the railroad (non-track) data.

On the positive side, to the extent that a causal relationship has been established from investment to patents, it is perhaps worth considering what kinds of factors may underpin the causal mechanism. The analysis of Chapter 3 implies that the social value of an invention and the private value of a patent on that invention are proportional to the extent of its application. With this in mind, and taking into account the fact that inventive activity itself, as well as patenting, are forms of investment, it is possible to see invention being 'caused' by capital goods investment through a version of the accelerator process. Most accelerator theories of investment rely on a form of naive expectations formation on the part of investors. The future is extrapolated from recent changes in the past. In the same way it could be argued that potential inventors or patentees, noting recent and current changes in the demand price of their inventions, extrapolate these changes to the future period over which the patented invention will yield returns. As noted above, the demand price of an invention is proportional to the potential extent of its application which, for capital goods inventions, is well proxied by gross capital formation. It follows therefore that inventive activity may be spurred by the observation of increasing investment in plant and machinery in an industry.

One of the difficulties with the foregoing tentative explanation of the causal mechanism is its reliance, in common with other accelerator theories, on an *ad hoc* and naive version of expectations formation. Expectations are of course unobservable, so speculation as to how they might be formed is not likely to illuminate the causal mechanism. Some mild support for viewing invention and patenting as a form of investment can, however, be found in the role of the concomitant variables in the regressions. It can be argued for example that there are certain pervasive common influences on all forms of investment. The cost of finance in the form of interest rates is one and the state of business confidence another. It follows therefore that when investment in railroads is regressed on past and future patents in railroads, the significance of the future patent variables should be influenced by the inclusion or otherwise of investment in manufacturing industry as a concomitant variable. When that variable is included its effect should be to reduce the influence of the common investment factors in the causality relation because the dependent variable is now effectively that part of railroad investment that cannot be 'explained' by contemporaneous manufacturing investment. Thus if invention is a form of investment that is affected not only by its specific expected returns (accelerator) but also by costs (rate of interest) and other common factors, regressions that exclude these factors will attribute that part of causation to the dependent variable and hence tend to increase the significance of the future patent variables. Examination of the role of manufacturing investment as a proxy for these common influences in Tables 6.4 to 6.6 confirms that it has the expected effect.

Finally, it is important to repeat the caveat mentioned in section 6.4 that the concept of Granger causality relates to the statistical predictability of one time-series from another, and nothing more. It follows therefore that to infer true causality from this statistical definition requires in addition a *post hoc ergo propter hoc* argument and a conviction that all relevant factors have been taken into account so that spurious correlations are avoided. An example of such a spurious attribution of causality would occur if both investment and invention are affected by the state of business confidence, which is not observed, and if there were a slower response of invention then it might appear statistically that invention is caused by investment whereas in truth both are caused by the unobserved variable. It seems therefore preferable not to regard the empirical results of this chapter in isolation but rather as a form of circumstantial evidence, to be taken together with different independent sources of evidence to be found in the following chapters.

NOTES

1. Schumpeter, of course, emphasised the act of entrepreneurship which, when associated with an invention, might give rise to an innovation. Innovations might not involve any invention at all, but if they do it is as an exogenous and important auxiliary factor.
2. Quoted in Morton Kamien and Nancy Schwartz (1982).
3. From a book entitled *Inventors and Money Makers*, quoted in Kamien and Schwartz, op. cit., p. 7.
4. See the compilation of his data published posthumously (Schmookler, 1972).
5. In the empirical work reported in section 6.5 the two series are spliced together by estimating a simple logarithmic regression between them for the 1870–92 overlap period, and then estimating gross capital expenditures for 1835–89 from that regression function. The estimated equation is:

$$\ln (\text{investment}) = 0.6943 + 0.6497 \ln(\text{miles})$$
$$(0.1181)$$

   $R^2 = 0.59$.

6. A moving average preserves the sums of the series within the span of the smoother whereas a running median will reflect 'typical' values and heavily discount aberrant points. Which of these, or any other likely criterion, should be preferred from the point of view of a dating exercise is not unambiguous. It all depends on what the average is meant to represent.
7. A moving average of random data will appear to exhibit cycles. This was first observed by Eugen Slutsky (1927), and was used by Ragnar Frisch (1933) in an attempt to demonstrate that 'business cycles' are statistical artefacts.
8. This is the reason why Schmookler's experimentation with the price of railroad stocks, reported in 6.3, could be misleading. Such asset price series are often used nowadays as 'leading indicators' of related real variables. In the context that Schmookler used these data it probably biased his assessment of the number of patent leads downwards.
9. In fact, a practical procedure for assessing Granger causality was first proposed by Christopher Sims (1972), who notes that it rests on a sophisticated version of the *post hoc ergo propter hoc* proposition.

10. A 'portmanteau' χ-square test based on the sum of squared cross-correlation coefficients is often used, see for example Pierce and Haugh (1977).

11. This may be why Pierce and Haugh (1977) failed to detect any causal relation in Granger's sense between macroeconomic variables, which has given rise to some dissatisfaction with this approach. See the discussion by Sims and Granger following their article.

12. Sims (1977) claimed that the filter $(1 - 0.75L)^2$ would 'flatten most economic time-series'.

13. If it is assumed that the e_i in equation 6.1 are white noise disturbance terms, then the joint autoregressive system could be estimated consistently by two-stage least squares, treating the lagged xs and ys as predetermined variables. This implies that instruments are needed only for the current values x_t and y_t.

14. This is different from the direct test described in Harvey (1981) which excludes the current value of x_t. Harvey's procedure in effect derives the reduced form of 6.2 as the regression equations so that only predetermined values of y and x appear as regressors. Including the current value of the putative causal variable allows possible instantaneous causality to be detected though it reduces the efficiency of detecting unidirectional causality. This form of the direct test is comparable with the indirect test which includes the current value of the 'caused' variable.

15. The F statistic with (m, T-n-m-k) degrees of freedom is calculated as:

$$\frac{(SSR^\circ - SSR).(T - n - m - k)}{m.SSR}$$

in which SSR and SSR° are the sums of the squared residuals from the full regression and the regression excluding the lagged x values respectively, and T is the number of observations.

16. The small sample properties of the direct and indirect tests are examined in Guikey and Salemi (1982), who marginally prefer the direct test, in contrast to Geweke *et al.* (1982) who judge the indirect test to be superior.

17. With the exception of this variable, all the data are presented in Schmookler (1966). The real investment in manufacturing data is from *Historical Statistics of the US, Colonial Times to 1970*, series P111 (structures) and P112 (equipment).

18. This is not altogether surprising as the early parts of this series have a median patent count of only 13, and it is a variable that can in principle go to zero. In these circumstances the logarithmic transformation easily produces an unstable series, and the effect in the regressions would be to inflate the importance of these early observations. Alternative procedures for dealing with this problem are (i) to use a less severe transformation such as the square root, which is recommended by Tukey (1977) for count data, or (ii) to add an arbitrary variance-stabilising number before taking logarithms. The

former option was experimented with, but revealed the opposite form of time-dependent heteroskedasticity.

19. Of course, this procedure is *ad hoc*, and does not give valid test statistics because, for example, it is possible for the correlation matrix not to be non-negative definite (though this did not occur in the current application).

7 The Supply of Inventions as a Function of the Level of Activity

The aim of this chapter is to complement the analysis of Chapter 6 by investigating further the question of endogeneity of invention. This is done by examining the empirical relationship between the number of inventions and the size of the industry in which they might be applied. Here use is made of Schmookler's cross-section data for a cohort of industries over 14 census years. Section 7.2 recapitulates Schmookler's results, and provides a possible interpretation of his regressions, and 7.3 re-examines these relations with a more explicit model and, it is argued, a more appropriate statistical method.

7.1 INTRODUCTION

It will be recalled from the theory presented in Chapter 3 on the incentives to invent that the derived demand for invention is a function of industry size. The theory was presented in terms of the incremental value to inventions of varying size, measured by unit cost reduction. It is straightforward to interpret the theory in terms of the number of inventions of a given size, or the equivalent number of average-sized inventions, if it is assumed that there are no interactions between inventions so that the total value of a set of inventions is equal to the sum of their individual values. The theory of the derived demand for inventions is therefore capable of providing a connection between the level of inventive activity and the level of economic activity. In other words, the theory offers scope for parameterising the observed statistical relationship reported in the Chapter 6. Recall that the causality analysis itself is essentially non-parametric.

The aim of this chapter is to propose a theoretical interpretation of the relationship between economic activity and inventive activity, and to estimate the parameters of interest in that relationship. One possible approach in the estimation phase would be to return to the annual time-series of the previous chapter and now estimate a transfer function formulation of the relationship in which possibly the long- or short-run multipliers bear a useful interpretation. There are two reasons why this path is not followed. First, the annual time-series data relate to about one century in history and it is implausible that a stable dynamic structure would apply for the whole span of time. Indeed, a number of authors have pointed to variations in the 'propensity to patent' and also variations in administrative lags in the patenting process over this period of time. But if the whole period is divided into sub-periods it becomes less feasible to apply transfer-function methods because of data demands. Unambiguous results from time-series analysis often require a large number of data points. Secondly, Schmookler himself provides an alternative and possibly more fruitful dataset in the form of a sequence of industry cross-sections. It is this latter dataset that is used for the empirical content of this chapter. This chapter complements the analysis of Chapter 6, which examined the question whether invention is exogenous. Here, exogeneity can be interpreted as a zero elasticity of supply, so the aim is to estimate this elasticity.

Section 7.2 describes Schmookler's data, summarises his analysis, and offers an interpretation of his results. It is argued that the relationship investigated by Schmookler is a supply curve for counts of inventions. Section 7.3 addresses some reservations about the way in which Schmookler carried out his analysis and presents the results of a re-examination of his data. A modified statistical model is proposed which implies a random coefficient estimation procedure and additionally introduces two extra variables to account for shifts over time in the supply schedules and in the relative profitability of invention in particular industrial applications. The estimation results confirm the responsiveness of inventive supply to economic forces, though the elasticity of 'supply' is less than unity. There is also evidence of autonomous supply factors and of the substitutability of inventive activity between fields of application.

7.2   SCHMOOKLER'S CROSS-SECTION/ TIME-SERIES ANALYSIS

In Chapter 7 of *Invention and Economic Growth*, Jacob Schmookler develops his previous joint notes with Z. Griliches

(1963) and O. H. Brownlee (1962) which are addressed to the question: how do investment and capital goods invention compare across a selection of industries? Having limited data on investment across industries — the US Census of Manufactures only providing such data from 1939, giving him just two observation periods consistent with his invention statistics — he then turns to value-added as a proxy for the level of investment in order to allow comparisons going back to 1899.[1]

The two years for which investment figures are available do, however, produce very good fits to simple log-linear regressions of total patents in years $t + 1$ to $t + 3$ regressed on investment in year t. For both years the proportion of variance in (log) patents explained is over 90 per cent and the estimated elasticities are extremely significantly different from zero, but do not differ significantly from unity. Schmookler concludes that 'this implies that inventive activity with respect to capital goods tends to be distributed among industries about in proportion to the distribution of investment. To state the matter in other terms, a 1 per cent increase in investment tends to induce a 1 per cent increase in capital goods invention.' He continues by rejecting the possibility that the high correlations obtained are due to joint dependence of both variables on industry size, and further observes that the coefficient on investment remains significantly different from zero 'even when the number of antecedent patents is introduced as an independent variable in the equation for patents.'[2]

Turning now to that part of Schmookler's results in which value-added is used as a proxy for investment, as they are reported in his book these results are very similar to those mentioned above with the elasticity coefficient on the 'demand variable' (value-added) never differing significantly from unity in 16 cross-section regressions for Census of Manufactures years from 1899 to 1947. This is also true for the reported pooled regression in which all the data for all industries and time periods are combined. However, since the data consist of observations on number of patents and value-added for a cohort of industries in 16 observation periods, it is curious that Schmookler does not report in greater detail the results of simple time-series analysis for individual industries, except almost paranthetically in the last paragraph and in a condensed footnote reporting on variants of the pooled regression with dummy variables. The reason appears to be that these experiments resulted in much lower determinacy of the elasticity coefficient to which Schmookler attached great importance. This is taken up again in section 7.3 of this chapter.

Schmookler inferred from his regression results that invention is induced by economic activity in the industry that might use the invention, and with an elasticity of unity. Though this interpretation has been disputed,[3] it can in fact be strengthened and sharpened. It will be argued that the relationship actually identifies a supply curve of inventions, though the re-examination of the data reported in section 7.3 implies a supply elasticity for counts of inventions substantially less than unity.

Consider how to measure the output of inventive activity. The discussion in Chapter 3 provides some useful ideas. Inventions are heterogeneous, but it seems reasonable to assert that a set of inventions that would reduce unit costs in a particular industry at a given time by, say, 10 per cent represents twice as much inventive output as a set of inventions that would reduce unit costs in the same industry at the same time by 5 per cent. It does not seem possible to construct a measure of the output of inventive activity in this way that would apply across industries and over time. On the other hand, there is greater homogeneity if the measure of inventive activity is approached from the input side. Inventive resource inputs of labour, capital services, and so on are substitutable to quite a large degree between industries of application for the invention, and the character of these inventive inputs may change only slowly over time. Thus it is better to interpret Schmookler's regression as applying to the supply of inventive effort than to the supply of technical improvements.

The value of a given invention is proportional to the extent of its application, and it may be assumed that the extent of application can be roughly measured by the non-inventive output or resource inputs of the industry utilising the invention. Thus, an invention that reduces unit production costs by, say, 1 per cent in the coal-mining industry is twice as valuable as one that reduces production cost in the leather goods industry by 7 per cent since the value of factor inputs in coal-mining is 14 times that in leather goods. Combining the two notions – first, that the 'volume' of inventive output is measured by reduction in unit cost; and secondly, that its value is the product of 'volume' and the size of the industry – it can be inferred that the *unit value* of an invention at any time and for any invention-utilising industry is proportional to industry size. This unit value can be interpreted as the demand price of inventions, since only demand-side factors have been used in its formulation.

If the size of an industry measures the demand price of invention in that industry, and this price enters as a parameter in inventors' decisions, then Schmookler's correlations between patents and value-added identify an invention *supply* function directly. Schmookler's 16 census year cross-section regressions of log(patents) on log(value-added) for a selection of between 14 and 20 United States industries all give a slope coefficient insignificantly different from unity. This would imply a supply elasticity equal to one, though Schmookler himself did not express his results in quite these terms.

This interpretation of Schmookler's regressions would still be valid if the intercept were allowed to vary across industries while the slope is assumed constant. Thus all industries might be assumed to have the same elasticity of invention supply, but the scale factor could differ across industries. When the selected industries are treated as a sample from a wider population the variation of the intercept within the sample is indistinguishable from the usual additive error term, and so long as the scale variation for the invention supply function is uncorrelated with industry size there will be no bias in the estimated slope (elasticity) coefficient.

Figure 7.1 illustrates the foregoing argument. Five industries are assumed, each having a distinct invention supply curve but with a common elasticity. In case (a) there is only a slight correlation between 'inventive opportunity' (cf. Scherer, 1965) in different industries, indicated by the height of the invention supply curve, and the potential extent of application of inventions as given by some measure of industry size. In case (b) a high negative correlation is assumed. The observations are shown as circled points and the dashed lines represent the implied sample regression functions. The effect of the assumed negative correlation between inventive opportunity and industry size is to bias the estimated elasticity downwards. A positive correlation would have implied positive bias or overestimation of the elasticity.

Rosenberg (1974) asserted that, because the complexity of the production technology and the capacity of technical knowledge to solve problems related to it differ between industries, it must be expected that invention supply curves would also differ, reflecting

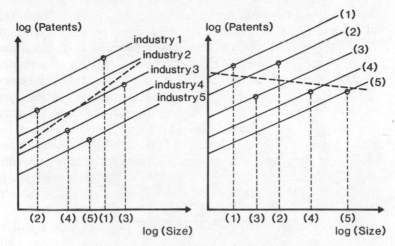

Figure 7.1: Bias in cross-section regressions

the differential costs of obtaining improvements. Now it can be seen that, of itself, this plausible assertion does not undermine the supply curve interpretation of Schmookler's regression results. To do so requires an additional assumption about how the supply curves differ: either that the elasticities of supply differ, or else that the scale factors are correlated with size. Thus Schmookler's cross-section regression results may be interpreted as supply curves if it is assumed that the elasticity of supply is constant across industries and that the industry shift factors are not correlated with size.

Unfortunately, there may be a case for arguing that the industry shift factors *are* correlated with size. Consideration of the 14 individual industries that make up Schmookler's basic cross-section shows that among the larger ones are: stone and clay products, petroleum refining, all other paper, all other textiles, and all other lumber and timber products. Among the smaller ones are: paper bags, envelopes, linoleum, cordage and twine, and cooperage. The larger industries tend to be conglomerates fabricating a range of products, whereas the smaller ones tend to be almost single-product industries. It seems quite probable therefore that there is an association between value-added and the number of distinct production technologies in Schmookler's sample so that his results could simply imply that the amount of invention is related to the scope for possible improvements in production technology. Figure

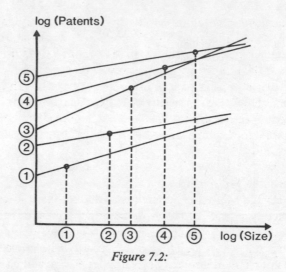

Figure 7.2:

7.2 illustrates the case in which there is a perfect rank correlation between the scale factors and industry size, though the elasticity of supply differs randomly across industries. It can be seen that the cross-section slope coefficient has a bias towards the regression coefficient of the intercepts on the size variable. For this reason the pure cross-section results which Schmookler emphasises must be treated with caution.

7.3   RE-EXAMINATION OF THE DATA

(i) The regression model

It was argued in the previous section that the invention supply function could be expressed $I_{it} = F(Q_{it})$ where I_{it} is the volume of inventions that can be applied in industry i at time t and Q_{it} is a measure of the size of the invention-using industry. When the volume of inventions is proxied by the count of successful patent applications ($I_{it} \sim P_{it}$) and the size of industry is taken to be valued-added ($Q_{it} = V_{it}$), and a constant elasticity form is assumed, then an estimating equation of the form

$$\log(P_{it}) = a_{it} + b_{it}.\log(V_{it}) + u_{it} \qquad 7.1$$

may be written, where u_{it} is a random error term. The double-subscripted parameters a_{it} and b_{it} indicate that both the scale factor and the elasticity might in principle vary across industries and over time. As it stands, the equation is impossible to estimate, requiring 2NT parameter values from only NT data points (N is number of industries, T is number of time-periods). To render the equation estimable it is necessary to place restrictions of some kind on the parameters. How such restrictions might be formulated is one theme of this section. The other theme is the adequacy of the set of regressors to capture the essence of the relationship under investigation.

(a) Parameter restrictions

Schmookler had himself experimented with various restrictions on the parameters a_{it} and b_{it}. His cross-sectional regressions for given time-periods imply $a_{it} = a_t$ and $b_{it} = b_t$ for example, and his pooled regression implies $a_{it} = a$ and $b_{it} = b$. Similarly, the pooled regression with intercept dummy variables implies $a_{it} = a_i$ and $b_{it} = b$, and so on. It was pointed out in the previous section, however, that the cross-section regressions, which Schmookler emphasised, should be treated with caution as they might be biased upwards. Schmookler's preference for them over the time-series regressions may have been largely due to their apparent consistency compared to the ambiguity in the time-series regression, but the consistency could be due to a systematic bias exhibited in the cross-section regressions, including the pooled regressions. The problem with the time-series regressions, which should be free of the bias, is that the variability of the estimates makes generalisation difficult.

Consider a different way of restricting the parameters a_{it} and b_{it}. By treating the set of industries as a sample from a wider set or population of industries, each with different values of the scale and elasticity parameters, it is possible to focus on the average or typical values that these parameters have in the population. In other words, consider the restrictions $a_{it} = a_i$ and $b_{it} = b_i$, where a_i and b_i are, for each industry, realisations from a random sampling process. There is a set of (a_i, b_i) pairs in the population of industries, and the actual sample represents a random sub-set of pairs from that population. The focus of interest is now on the population expected values for a_i and b_i, and these may be estimable from the data. With this formulation the problem is now one of 'random coefficients regression' as presented for example by Swamy (1970).

Suppose that for any given industry the regression coefficients are constant through time, but that they differ across industries because invention cost curves differ across industries. This would imply the following restrictions:

$$\alpha_{it} = \alpha_i, \quad \beta_{it} = \beta_i \qquad 7.2$$

and would justify separate time-series regressions for each industry. However, such a procedure would not easily allow general, as opposed to industry-specific, conclusions about the nature of the invention cost curve unless the coefficients were both well determined and very close across industries, in which case a pooled time-series and cross-section regression would be appropriate for the general relationship.

Now suppose that both common factors and industry-specific factors affect the coefficients in the relation between patents and value-added. Suppose further that the incidence of industry-specific factors is random, but drawn from a common distribution over time. Then the implicit homogeneity restrictions are:

$$\alpha_{it} = \alpha + \alpha_i, \quad \beta_{it} = \beta + \beta_i \qquad 7.3$$

where α_i and β_i are now random variables with zero means and constant variances and covariances, and α and β are population expected values so the specification conforms to that of the random coefficients regression model.

It is possible to estimate a variant of random coefficient regression in which a sub-set of the coefficients are deterministic and equal across industries while the rest of the coefficients vary randomly across industries. This is achieved by setting *a priori* the appropriate terms in the coefficient covariance matrix to zero.

(b) The set of regressors
In addition to the simple model:

$$\log(P_{it}) = \alpha_i + \beta_i.\log(V_{it}) + u_{it}, \qquad 7.4$$

where α_i and β_i are written for convenience instead of $\alpha + \alpha_i$ and $\beta + \beta_i$, three other regression equations were estimated by introducing two new independent variables with the following justification.

A problem of some importance regarding the simple model is that, interpreted as a supply curve, it applies at a particular point in time and, of course, with the usual *ceteris paribus* qualification, whereas the data span 51 years from 1900 to 1950. There is no doubt that changes of some importance took place over this period of time, affecting both 'supply' and 'demand' factors. Schmookler himself saw the need to include time as a separate independent variable in order to improve goodness of fit. But of itself a 'time' variable explains little and is not much more than an *ad hoc* fitting device. It is preferable to allow for shifts in 'supply' and 'demand' factors in a more explicit manner.

The slightest acquaintance with the recent history of science and technology would lead one to suppose that the first half of the twentieth century experienced a very rapid growth in the knowledge base from which inventions draw. This 'march of science' was emphasised by Rosenberg, and would, to a certain extent, be a common factor influencing the invention supply curves of various industries. Thus all invention supply curves might be expected to exhibit co-movements over time as a common knowledge base affects the cost of making individual inventive improvements. While this may be the most fundamental source of co-movement, there may also be other more prosaic, but none the less statistically important, common factors in the data such as variations in the propensity to patent for institutional causes. In order to take such effects into account, another variable was constructed, namely the logarithm of the number of patents filed in the United States in all areas other than capital inventions in the industry in question, $\log(TP_t - P_{it})$. It would be expected to have a coefficient equal to approximately one if the proportional impact of these common factors on the particular industry is roughly equal to their impact elsewhere. In other words, if the shift in the industry supply schedule is typical.

The other variable to be introduced is $\log(TV_t - V_{it})$, or total value-added in other industries. Its rationale is that if at any point in time there is a given amount of inventive resources to be allocated across the different industrial fields of application, then it might be expected that the number of patents filed in any field reflects not just the absolute size of the industry but also its relative size. If Schmookler is right in his argument that the scientific–technical knowledge base from which inventions draw is sufficiently wide and flexible that it can be used to create improvements in whatever area is desired, then there should be high substitutability on the supply side between inventive output in different fields. An expansion elsewhere would be expected to draw inventive resources away from the industry. It was argued in section 7.2 that the size of an industry measures the demand price of inventive improvements in that industry, and in this context it is natural to see it as a relative price. Thus the coefficient on this variable may be interpreted as a 'cross-elasticity' and may provide a further test of the influence of demand factors on inventive activity.

Including both variables gives a second estimating equation as:

$$\log(P_{it}) = \alpha_i + \beta_i.\log(V_{it}) + \gamma_i.\log(TP_t - P_{it}) + \delta_i.\log(TV_t - V_{it})$$
$$7.5$$

The final equation was estimated as a constrained version of equation 7.5. By constraining $\gamma_i = 1$ and $\delta_i = -\beta_i$, the following is derived:

$$\log(P_{it}/(TP_t - P_{it})) = \alpha_i + \beta_i.(\log(V_{it}/(TV_t - V_{it}))). \qquad 7.6$$

Here the variables are explicitly defined in relative terms. The ratios can be interpreted as giving the probabilistic odds of a particular patent (or dollar of value added) arising in industry i.

(ii) Data and results

The data used in the regressions are given in Schmookler, *Inventions and Economic Growth*, Chapter 7. The value-added measure for industry i in year t, V_{it}, is defined at constant 1926 prices using the BLS all commodity index of wholesale prices as the deflator. The index t relates to Census of Manufactures years running from 1899 to 1947, which had five-year intervals between 1899 and 1919, and two-year intervals from 1919 to 1939. There was one post-World War II observation (1947), making 16 observation periods in all.

The number of successful patent applications on capital goods inventions related to industry i for the years $t + 1$ to $t + 3$ is denoted P_{it}. These data were collected by Schmookler from US Patent Office records and an account of their derivation is given in Schmookler (op. cit., Chapter 1).

The estimate of the total number of successful patent applications filed at the United States Patent Office for the years t = 1 to t + 3, TP_t, is given in Schmookler (ibid., Table A4). Total value-added for all industries in the United States in year t at constant 1929 prices, TV_t, is from United States Department of Commerce (1975).

The results for each estimated equation are presented in two parts. First the equations estimated by ordinary least squares from a time series of 16 observations on each industry are given with a mnemonic label on the left to indicate the relevant industry. The first columns report the estimated coefficients of the equations with their standard errors in parentheses below. Next there are two columns reporting the 'goodness of fit', the standard error of the equation (SE) and the coefficient of determination uncorrected for degrees of freedom (R^2). Finally, the Durbin–Watson statistic is presented as a check on the presence of first-order serial correlation in the residuals.

The mnemonic labels for the industries are as follows:

SC: stone and clay products
PR: petroleum refining
FO: footwear except rubber
GL: glass
PG: paper bags
EN: envelopes
PX: paper boxes
OP: all other paper
LI: linoleum
CR: other carpet and rug
CT: cordage and twine
OT: all other textiles
CO: cooperage
LT: all other lumber and timber products

After the detailed time-series regression results for each industry is a presentation of random coefficient regression analysis for the pooled time-series and cross-section data. The X-square statistic tests the hypothesis of coefficient homogeneity. Then are given estimates of the mean coefficient vector, with standard errors, for different combinations of assumptions about the nature of individual coefficients. Thus certain coefficients might be assumed to be the same for all industries whereas others may differ ('randomly') across industries but be drawn from the same distribution. For the latter it is necessary to report not only the estimated mean value of the distribution with its standard error but also the estimated standard deviation of that distribution. The numbers in parentheses in the α_i, β_i, etc. columns represent these estimated standard deviations. If no number appears in the parentheses, that signifies that the coefficient has been taken to be the same in all industries (i.e. that it is not random, or that it has variance zero).

Discussion

The X-square tests for homogeneity of the coefficients are all very high and suggest a definite rejection of the hypothesis that the coefficients for different industries are equal. This implies that pooling the time-series and cross-section data in a simple regression is an incorrect method of aggregating the data.

For the individual industry regressions of equation 7.4 (Table 7.1), none of the goodness of fit statistics is 'satisfactory' and the equation coefficients are poorly determined. The random coefficient analysis of this equation is unable to detect with any precision an aggregate elasticity term. If the scaling term α_i is allowed to vary across industries but the elasticity β_i is assumed the same for all industries, the elasticity is not significantly different from zero. If both coefficients are allowed to differ across industries then, although the 'common effect' elasticity is estimated at 0.46, a 95 per cent confidence interval includes both 0 and 1. This gives an important negative conclusion for the basic specification which is diametrically opposed to the conclusion claimed by Schmookler. In other words, the evidence here does not appear consistent with the view that inventive activity responds to economic factors. However, if neither α_i nor β_i is allowed to vary across industries, the elasticity coefficient appears to be significant at 0.5 (it is also significantly different from one), and this inappropriate specification is similar to Schmookler's pooled regression except that the industries are combined through a generalised (variance-weighted) least squares analysis.

In equation 7.5 (Table 7.2) there is a remarkable improvement in the goodness of fit of the individual OLS equations due to the

Table 7.1

Equation 7.4: $\ln(P_{it}) = \alpha_i + \beta_i \ln(V_{it})$

i	α	β	SE	R^2	DW
SC	6.32	−0.27	.36	0.00	0.92
	(1.46)	(0.23)			
PR	0.69	1.00	.45	0.84	1.01
	(0.67)	(0.12)			
GL	4.75	0.28	.44	0.09	0.87
	(1.28)	(0.24)			
PG	2.90	0.29	.55	0.16	0.54
	(0.53)	(0.18)			
EN	3.61	−0.08	.37	0.02	0.57
	(0.45)	(0.15)			
PX	4.30	0.17	.17	0.36	1.03
	(0.29)	(0.06)			
OP	5.26	0.19	.38	0.10	0.25
	(0.90)	(0.15)			
LI	2.31	0.27	.47	0.18	0.75
	(0.50)	(0.15)			
CR	2.23	0.46	.44	0.18	0.84
	(1.10)	(0.26)			
CT	2.13	0.09	.44	0.00	0.98
	(1.50)	(0.44)			
OT	4.98	0.37	.16	0.45	0.69
	(0.81)	(0.11)			
CO	−7.33	3.17	.94	0.28	0.72
	(4.29)	(1.37)			
LT	6.57	0.02	.38	0.00	0.15
	(1.80)	(0.26)			

Random coefficient analysis of above equations: $\chi^2(24) = 650.2$

α	α_i	β	β_i	
(i)	3.42		0.46	
	(0.96)	(3.15)	(0.26)	(0.75)
(ii)	3.37		0.29	
	(0.32)	( )	(0.31)	(0.75)
(iii)	3.45		0.18	
	(0.95)	(3.15)	(0.13)	( )
(iv)	3.12		0.50	
	(0.31)	( )	(0.05)	( )

Table 7.2

Equation 7.5: $\ln(P_{it}) = \alpha_i + \beta_i \ln(V_{it}) + \gamma_i \ln(TP_t - P_{it}) + \delta_i \ln(TV_t - V_{it})$

i	α	β	γ	δ	SE	R^2	DW
SC	−1.97	0.37	1.35	−0.88	.25	0.59	1.62
	(4.76)	(0.36)	(0.38)	(0.41)			
PR	−0.39	1.34	1.19	−1.30	.39	0.89	2.02
	(9.16)	(0.28)	(0.57)	(0.79)			
FO	4.10	0.94	0.55	−0.84	.25	0.42	1.35
	(5.18)	(0.55)	(0.39)	(0.47)			
GL	−6.64	1.14	1.91	−1.37	.38	0.41	1.28
	(6.13)	(1.01)	(0.89)	(1.43)			
PG	20.74	1.01	0.01	−1.76	.52	0.37	0.70
	(8.98)	(0.64)	(1.21)	(1.56)			
EN	−0.67	0.50	1.62	−1.45	.22	0.71	2.39
	(5.12)	(0.26)	(0.32)	(0.48)			
PX	1.91	0.46	0.78	−0.72	.13	0.68	1.65
	(3.09)	(0.21)	(0.23)	(0.44)			
OP	−2.29	1.43	2.63	−2.72	.20	0.78	1.35
	(3.71)	(0.36)	(0.45)	(0.73)			
LI	13.87	0.92	0.53	−1.75	.42	0.45	1.22
	(9.21)	(0.32)	(0.61)	(0.73)			
CR	−13.87	0.23	1.29	0.17	.38	0.50	1.19
	(6.11)	(0.56)	(0.64)	(0.75)			
CT	−9.18	−0.71	0.75	0.46	.44	0.15	0.98
	(7.99)	(1.12)	(0.64)	(0.83)			
OT	4.05	0.73	0.37	−0.54	.13	0.71	1.14
	(2.58)	(0.24)	(0.20)	(0.24)			
CO	−2.25	0.17	3.30	−3.04	.37	0.90	2.29
	(7.28)	(0.70)	(0.55)	(0.37)			
LT	0.73	0.30	1.35	−1.05	.17	0.83	1.27
	(2.66)	(0.13)	(0.25)	(0.15)			

Random coefficient analysis: $\chi^2(52) = 764.9$

	α	α_i	β	β_i	γ	$.\gamma_i$	δ	δ_i
(i)	0.01		0.57		1.27		−1.12	
	(2.16)	(6.05)	(0.20)	(0.18)	(0.26)	(0.68)	(0.26)	(0.57)
(ii)	0.53		0.55		1.22		−1.10	
	(1.17)	( )	(0.20)	(0.18)	(0.29)	(0.68)	(0.28)	(0.57)
(iii)	0.09		0.71		1.30		−1.17	
	(2.13)	(6.05)	(0.16)	( )	(0.26)	(0.68)	(0.24)	(0.57)
(iv)	−0.22		0.54		1.18		−1.12	
	(2.15)	(6.05)	(0.21)	(0.18)	(0.20)	( )	(0.29)	(0.57)
(v)	−0.51		0.55		1.23		−0.98	
	(2.14)	(6.05)	(0.18)	(0.18)	(0.30)	(0.68)	(0.20)	( )

Table 7.3

Equation 7.6: $\ln\{P_{it}/(TP_t - P_{it})\} = \alpha_i + \beta_i \ln\{V_{it}/(TV_t - V_{it})\}$

i	α	β	SE	R^2	DW
SC	−3.31	0.42	.29	0.07	1.06
	(2.02)	(0.40)			
PR	2.47	1.38	.36	0.87	1.95
	(0.80)	(0.14)			
FO	−1.22	0.73	.24	0.16	1.19
	(2.41)	(0.44)			
GL	−2.53	0.49	.37	0.05	0.97
	(3.30)	(0.54)			
PG	−2.15	0.69	.60	0.25	0.60
	(2.72)	(0.32)			
EN	−9.78	−0.17	.30	0.04	1.11
	(1.93)	(0.23)			
PX	−5.07	0.23	.14	0.29	1.24
	(0.65)	(0.10)			
OP	−3.72	0.30	.27	0.14	0.34
	(10.9)	(0.21)			
LT	−5.13	0.42	.46	0.19	0.10
	(1.90)	(0.23)			
CR	−5.66	0.27	.39	0.02	0.98
	(5.49)	(0.49)			
CT	−10.59	−0.17	.41	0.00	0.90
	(5.38)	(0.68)			
OT	−2.40	0.41	.16	0.14	0.54
	(1.07)	(0.27)			
CO	5.79	1.82	.59	0.71	1.24
	(2.57)	(0.31)			
LT	−2.34	0.61	.27	0.46	1.23
	(0.79)	(0.18)			

Random coefficient analysis: $\chi^2(26) = 552.9$

	α	α_i	β	β_i
(i)	−2.84		0.58	
	(1.06)	(3.4)	(0.16)	(0.39)
(ii)	−2.54		0.64	
	(0.39)	( )	(0.22)	(0.39)
(iii)	−2.96		0.55	
	(1.06)	(3.4)	(0.19)	( )
(iv)	−1.22		0.83	
	(0.24)	( )	(0.04)	( )

inclusion of 'supply factors' (TP_t-P_{it}) and 'substitution factors' (TV_t-V_{it}). The Durbin–Watson statistic has also improved considerably, though it still indicates significant positive autocorrelation in most cases. This could be a signal of mis-specification, but could equally reflect the fact observed by Schmookler that the 'last four observations are drawn from the depressed 1930s when normal investment value-added ratios were badly disturbed'. The hypothesis of coefficient homogeneity across the OLS equations is again decisively rejected by the χ^2 test. When the OLS equations are aggregated to form the random coefficient estimator, the elasticity of supply is estimated at 0.57 and is significantly different from both 0 and 1 at the 95 per cent confidence level. It is noteworthy that the elasticities γ and δ are relatively well determined and do not differ significantly from 1 and -1 respectively, although there is considerable variability in the individual equations. As far as γ is concerned, this implies that a 1 per cent increase in patents elsewhere in the economy is on average associated with a 1 percent increase in patents for the industries examined, *ceteris paribus*. This appears to vindicate the view that autonomous supply factors are important. On the other hand, the estimated coefficient δ implies that 1 per cent increase in value-added elsewhere in the economy should lead, *ceteris paribus*, to a 1 per cent decline in patents in the industries examined. This, together with the significance of the elasticity β, is consistent with the view that economic factors are also relevant to the supply of inventive effort generally.

Not surprisingly, equation 7.6 (Table 7.3) which is an *a priori* constrained version of equation 7.5, does not perform well in comparison with that equation. However, it represents a considerable improvement in goodness-of-fit terms over the basic model represented by equation 7.4.

Another constrained regression equation was derived from equation 7.5 by suppressing the scale intercept term which had been insignificant. The results were broadly similar to those for equation 7.5, and have not been tabulated.

7.4 CONCLUSION

This chapter has provided an interpretation, a reformulation and a re-estimation of the relationship between industry size and the number of inventions relating to that industry. Pulling the strands together, it appears that there is strong evidence that economic factors do influence the level of inventive activity. This strengthens the conclusion of Chapter 6 with regard to the endogeneity of inventive activity. However, it also appears to be the case that the elasticity of invention supply is not as high as unity. It probably lies between $\frac{1}{2}$ and $\frac{3}{4}$ for a 'typical' US industry in the first half of the century, and it should be noted that there is considerable variability between industries. The ability of demand conditions elsewhere in the economy to pull inventive resources away from an industry also appears to be quite strong for the 'typical' industry, but highly variable between industries. This, too, corroborates the endogenous invention proposition.

NOTES

1. The set of industries was: petroleum refining; synthetic fibres; glass; sugar refining; tobacco manufacturing; railroads*: stone and clay products; knitting*; narrow fabrics; mats; linoleum; other carpets and rugs; cordage and twine; yarn and thread*; all other textiles; veneer mills*; cooperage; other lumber and timber products; pulp, paperboard and pulp goods*; envelopes; paper bags; paper boxes; all other paper.

   The five asterisked industries were not included in the value-added-based data, but the following were included (though not in the investment data): footwear except rubber; seamless hosiery.

2. In fact, these regressions including a lagged dependent variable give much stronger support to Schmookler's views than he apparently realised. If they are interpreted as the reduced form of a dynamic partial adjustment model, the implied equilibrium (long-run) elasticity is unity, which is consistent with his assertion in the preceding quotation.

3. In particular by Nathan Rosenberg (1974) in a major review article of Schmookler's book. Rosenberg attempts to reassert the importance of autonomous supply factors. It is possible to reconcile Schmookler's view

with Rosenberg's if it is understood that the former relates to the supply of inventive effort while the latter is about technological improvements. The importance of this distinction is highlighted by the observations in Chapter 8.

8 The Supply of Inventions: Further Considerations and Analysis

8.1 INTRODUCTION: A TAXONOMY FOR THE SUPPLY OF INVENTIONS

In an article entitled 'The Supply of Inventors and Inventions' (1962), Fritz Machlup gave an overview of the issues from an *a priori* standpoint. Starting with the statement:

The analysis of the supply of inventions divides itself logically into three sections: (1) *the supply of inventive labor* – the chief input for the production of inventions; (2) *the input–output relationship* – the technical production function describing the transformation of inventive labor into useful inventions; and (3) the *supply and cost of useful inventions* – the output obtained from the use of inventive labor. All this, of course, follows the pattern by which the supply of any economic good is analysed in modern economic theory.

Machlup concluded with four elasticity propositions:

(1) The supply of inventive labor is unlikely to be infinitely elastic and quite likely to be relatively inelastic; (2) The supply of inventive labor capacity is probably even less elastic than the supply of inventive labor; (3) The supply of new raw inventions may, in certain circumstances, be even less elastic than the supply of inventive labor capacity; and (4) The supply of effective (worked) inventions is likely to be even less elastic than the supply of raw inventions.

These decreasing elasticities at successive stages of the inventive process were ascribed to what Machlup called:

four potential shrinkages in the percentage increase in yield: higher rates of pay, lower quality of the personnel, smaller output of raw inventions per input of inventive capacity, and a higher rate of rejection in the selection of inventions for use.

He went on to state that: 'These shrinkages are independent of one another; but they may add up with a vengeance.' As noted, Machlup's conclusions rest largely on *a priori* reasoning. Empirical evidence has in the past been sparse, although the situation has

improved in recent years with the efforts of researchers such as
Schmookler and Mansfield and with the initiation of official
surveys of R & D. The aim in this chapter is to examine the sources
of diminishing returns, or supply inelasticity, from an empirical
standpoint.

By supposing inventor-entrepreneurs to be 'economic men',
maximising a utility or profit function subject to resource constraints, it becomes possible to derive a supply function and,
equivalently, factor demand functions. In fact, the whole concept
of a supply function is contingent on the assumption of economic
rationality. But, in addition to allowing one to talk in such terminology, the rationality assumption, with its implications of cost-
minimisation, points the way to unravelling the mechanisms that
underlie the supply function. This is Machlup's essential insight. It
permits the applied researcher to approach the supply function not
only directly as in the previous chapter, but also indirectly. In other
words, to establish a relationship between output and inputs is
equivalent to establishing the supply curve. This follows from the
theory of duality (see Shephard, 1970; Varian, 1980).

The structure of this chapter follows Machlup's classification of
the sources of supply inelasticity, but not in the order in which he
presented them. The next two sections take up Machlup's important observation that if it is possible to order the agenda of research
projects according to expected profitability, this itself is a potential
source of diminishing returns. First, in section 8.2, the idea is
grafted onto the supply of counts of inventions, a topic that was
treated in the previous chapter, in order to derive the implications
for the supply of new technology implicit in an estimate of the
elasticity of supply of the number of inventions. Next, in section
8.3, it is argued that it may be possible to infer something about
the elasticity of supply of effective invention directly from the
distribution of inventions by value. The indirect approach to supply, via the transformation of inputs into outputs of inventions is examined in sections 8.4 and 8.5. Following on from the ideas
presented in section 8.3, it is argued in 8.4 that the distribution of
researchers by productivity has certain implications for the supply
function. In section 8.5 the indirect 'production function'
approach is examined in an international inter-industry macro
setting. Finally the conclusions presented in section 8.6 suggest that
all the pointers are in one direction: the supply of inventions and

of new technology are influenced by economic factors, but the elasticity may in practice be rather low.

8.2   THE IMPLICATIONS OF FILTERING FOR THE SUPPLY OF NEW TECHNOLOGY

The previous chapter took for granted the idea that the number of inventions can be used as a measure of the level of inventive activity and inventive output. It was argued that the clearly established statistical relationship between the number of patents relating to capital equipment for use in particular industries and the size of those industries traces out a supply curve of patented inventions. It was noted, however, that Schmookler's analysis of his own data, which under this interpretation would imply a supply elasticity of approximately unity for counts of inventions, could be improved by more appropriate specification of the regression analysis. A tentative conclusion that the elasticty is nearer to 0.5 than to 1 was drawn – indeed, in the preferred equation a 95 per cent confidence interval ranges from 0.25 to 1.

However, there may be a problem in using counts of inventions (patents) as a measure of the output of inventive activity. It could be argued that the counts should somehow be weighted by the quality of inventions. After all inventions are by definition hetero-geneous, and in fact vary enormously in value. If the ith invention is indexed by a measure of quality, u_i, then the appropriate measure of inventive output would be:

$$I = \sum_{i=1}^{N} u_i$$

Now inventive output can be measured by the count N, if N is proportional to I. This would be true on the average if researchers are unable to distinguish the quality of their inventions before making them, so that the expected value of each invention appears to them to be the same. However, if researchers are able to discriminate between inventions by quality or net value then they might be expected to work through their agenda of research projects in order of profitability. The two situations might be pictured as in Figure 8.1

Hence this discriminating or filtering process implies that the

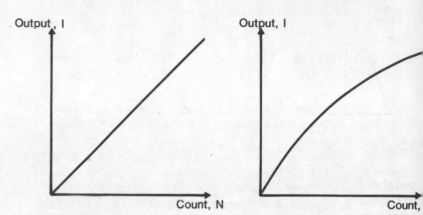

Figure 8.1: Inventive output and a count of inventions

elasticity of supply of inventions weighted by quality is less than that of pure counts of inventions. Bearing this in mind, the elasticities quoted above should be interpreted as upper bounds since they are based on an implicit assumption of no discrimination between inventions by inventive researchers.

The sense in which researchers 'work through' their agenda of projects should be understood as applying at an instant of time rather than over time. It means that research resources are allocated across the set of identified research projects so as to maximise the net value of the research resources. Of course, the value of a project depends on the set of cooperating technologies, which change over time.

There appears now to be a self-contradiction in the argument associating the statistical evidence of a relation between numbers of inventions and size of the potential invention-using industry with a supply function for inventions. While that argument relies on an assumption that inventors can discern profitable lines of enquiry and channel their resources in such directions, the use of unweighted counts of inventions implicitly assumes their inability to discriminate among research projects. This self-contradiction can be resolved if it is assumed that, within any application, there may be an *ex-post* distribution of projects by actual values, but *ex-ante* all projects appear alike in profitability so that the unweighted

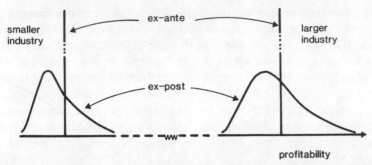

Figure 8.2: Ex-ante *and* ex-post *distributions of inventions by profitability*

count of inventions is proportional to the expected value of inventive output. But, at the same time, between industries of application inventors recognise that the size of the invention-using industry directly and proportionately affects the value of each and every invention so that the whole *ex-post* distribution is shifted, and the *ex-ante* expectation (mean) along with it. Thus consider the set of invention projects applicable to a particular industry at a point in time, and hypothesise two different sizes for that industry. The foregoing assumptions are reflected in Figure 8.2 which shows the *ex-ante* and *ex-post* frequency distributions of the set of projects by net value.

8.3 THE DISTRIBUTION OF INVENTIONS BY VALUE

Machlup (1962) asserted that diminishing returns are always due to the presence of some fixed factor, and named 'the existing stock of scientific knowledge and the state of the industrial arts at any moment of time' as one fixed factor, and 'the stock of known problems' as another. He also spelled out two independent sources of diminishing returns arising from a given stock of known problems. First, there is duplication of effort; and secondly, with a given agenda of problems, one can think of inventive activity proceeding by attacking what are thought to be the easier or more profitable problems first, and giving a lower priority to the more difficult or less profitable ones. This is the filtering process alluded to in section 8.2, and it continues as the main theme here.

If inventors discriminate between projects by expected profitability, then the distribution of projects by expected profitability will give the supply of inventions. Suppose that the inventor-entrepreneurs apply some external test to the profitability of their activity, where the test might be thought of as the opportunity cost of the resources used up by doing the research and development. Assuming that inventors require a given rate of return to engage in inventive activity, they will produce all those inventions that exceed that rate of return according to their expectations. In other words, the inventions supply curve will correspond to the cumulative frequency function derived from the expected profitability frequency distribution.

To interpret the cumulative distribution of research projects by profitability (ordered from the most profitable to the least) as a supply curve, observe that the profitability of all projects increases as the value or the price of a unit of inventive output increases. Hence the profitability cut-off given by the external test moves leftwards in Figure 8.3 (a) as the price of inventive output rises. This implies that the supply price is inversely related to the required rate of return, so a cumulative frequency function can be depicted for the set of research projects at varying unit prices of inventive output as shown in Figure 8.3(b). Simply interchanging the axes gives the required supply function.

Now suppose that the contrary case to that just considered holds true. That is, inventors cannot anticipate the profitability of their possible invention projects; they cannot distinguish, *ex ante*, good

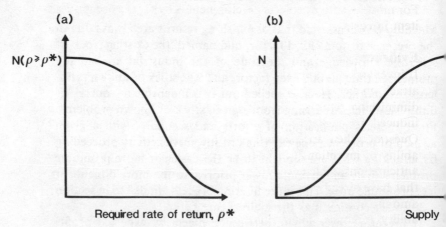

Figure 8.3: The distribution of invention projects by profitability

projects from bad. However they may be aware of the profitability of invention in general, so that on average they may expect an overall rate of return to obtain even though they cannot associate particular projects with any measure of profitability. Suppose that inventors are expected value-maximisers.[1] Then at any price lower than that corresponding to the overall average rate of return, no inventions will be supplied while at any higher price all available inventive effort will be put to use supplying inventions, which differ in value and profitability of course. Considered individually some will make losses but these will be counterbalanced by those that are profitable. The supply of inventions will be infinitely elastic at a supply price corresponding to the overall average expected rate of return.

It can be seen then that the ability to discriminate between good and bad invention projects makes the number of inventions supplied respond more inelastically with respect to price than would otherwise be the case. It also turns invention into a more profitable activity, considered as a whole. On this basis one can predict a derived demand for a filtering service that enables a better discrimination between potential invention projects. It will pay inventors to invest resources in screening.[2] And it is arguable that part of what is defined as 'applied research' in national surveys of research and development is, in fact, devoted to such screening.[3] This would make sense in terms of the typical commitment of resources to development as opposed to basic or applied research. For most inventive projects development is by far the preponderant item in costs.

Evidence

If inventors can order projects by potential profitability, a rational attack on the set of problems at hand will give rise to a form of diminishing returns, and a greater number of inventions will be induced by higher unit rewards. This is the supply function. Questions then arise concerning the degree to which the profitability of inventions can be foreseen, and the distribution of such anticipations, which are obviously empirical issues. For inventions that have already reached the development stage there is a certain amount of information available regarding two aspects of profitability forecasts, namely the estimated cost of the final article and the time needed for completion of development projects. The most well-known such estimates, and analysis thereof, relating to American military aircraft development, is that carried out by researchers at the Rand Corporation in the early 1960s (see Marshak *et al*; 1967). However, there has also been more recent confirmation of these results for firms in the non-aircraft business enterprise sector.[4] The main features of these data that have been observed are first, the tendency for *ex-ante* estimates of production cost to be much smaller than the outcome (the average 'cost factor'

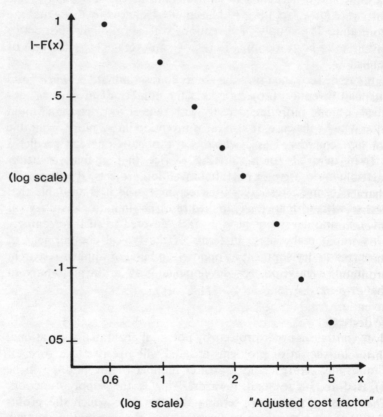

Figure 8.4: Distribution of cost factors

Note: The cost factor is the ratio of estimated cost to actual (out-turn) cost. The adjustments, carried out by Summers, relate to deflation for changes in price levels and to adjustment for learning curve effects due to procurement-quantity differences between the estimate and the outturn.
   Estimated value of α (exponent of the Pareto distribution): 1.3.
Source: T. Marshak *et al.* (1967).

– ratio of out-turn to estimate – is typically between 2 and 5); and secondly, the very wide variations in cost factors.

Clearly, for many invention projects, and especially for radical changes as opposed to minor improvements, the *ex-ante* guesses about profitability in an absolute sense are likely to be wild. However, the careful statistical analysis by Summers (in Marshak *et al.*, 1967) which examines these data demonstrates that if the initial estimates for two development projects stand in as low a ratio as 5:3 then what appears initially to be the more costly will turn out so four times out of five. It appears then that, at a given point in time, and faced with a particular set of potential projects, inventors *are* able to screen projects to some degree and so separate them one from another as regards profitability. The particular value of this evidence for the argument in this chapter is in the shape of the distribution of cost factors. The distribution displays a remarkably fat tail. In fact, a visual test suggests that the class of stable Pareto distributions may not be an unsuitable candidate for its form.[5] The data are pictured in Figure 8.4.

If the cost factors have a Pareto distribution, it may not be farfetched to suppose that the same kind of distribution may characterise the *ex-ante* expectations of profitability from inventive activity which, it was argued earlier, determines the shape of the invention supply curve. Of course, expectations about profitability are not easily observable, but if they are rational or well founded, they should be reflected in the actual, *ex-post*, distribution of profitability of inventive projects. Now, there is some evidence on the *ex-post* distribution by value or profitability of successful inventions, and it does accord with the supposition that they might be described by a Pareto distribution. Since the data are not extensive, it is possible to examine briefly all that are available, in chronological order of publication.

(i) B. Sanders (1958)
The results of this survey were noted by F. M. Sherer (1965) to be distributed *à la* Pareto – from which Sherer concludes for his own study that 'it forces us to acknowledge that patent statistics are likely to measure run-of-the-mill industrial inventive output much more accurately than they reflect the occasional strategic invention which opens up new markets and new technologies. The latter most probably remain the domain of economic historians.'[6]

Of course, the Pareto description only applies to those inventions that made a net gain. The Sanders survey indicates however that about one half of the 281 patents reporting a numerical value were loss-making. The losers are much more tightly packed around the bread-even median, however. The data are shown in Figure 8.5.

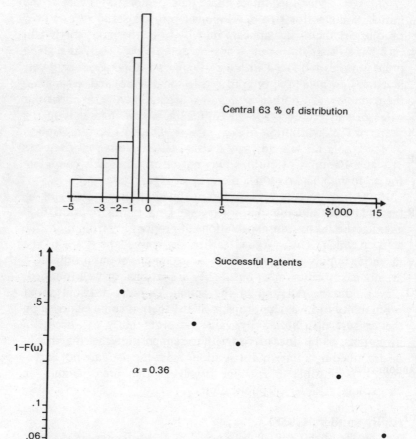

Figure 8.5: The distribution of patents by value

Note: The large proportion of patents (about one-fifth) that reported no gain and no loss were allocated equally to adjacent cells.
Source: B. S. Sanders *et al.* (1958).

(ii) J. L. Enos (1962)

Enos presented data for the profitability of inventions in the petroleum refinery industry. Since his data were presented as the personal returns to individual inventors, of whom more than half in his sample were salaried and presumably assigned their inventions to their employers, only five data points remain. These span a period of almost 30 years, over which time the output of the industry was growing strongly, and presumably the demand price for such inventions along with it. Even so, the impression that the cumulative distribution is linear on the log-log scale would remain if adjustment were made for the varying demand price.

It must be pointed out that Enos's sample is in no sense random. The inventions that he selected for scrutiny were among the more dramatic in the industry. He was obviously examining the upper tail of the distribution.

(iii) E. Mansfield *et al.* (1977)

Mansfield and his collaborators attempted to measure both the private and social rates of return from 17 industrial inventions that were put into production. Of these innovations, four were new producer goods and three were new consumer goods. They occurred in a 'wide variety of industries, and in firms of quite different sizes'. The authors go on to say that most were of 'average or routine importance, not major breakthroughs', and 'although the sample cannot be regarded as randomly selected, there is no obvious indication that it is biased toward very profitable inventions (socially or privately) or relatively unprofitable ones'. For 9 of the 17 innovations the authors state that they obtained:

data concerning the approximate private rate of return *expected* from the innovation by the innovator when it began the project.... In 5 of the 9 cases, this expected private rate of return was less than 15 per cent (before taxes), which indicates that these five projects were quite marginal from the point of view of the firm.... Yet the average social rate of return from these 5 innovations was over 100 per cent.... Among the innovations for which we have data, there is no significant correlation between an innovation's expected private rate of return and its social rate of return.

One is much less inclined to view these data as deriving from a Pareto distribution: possibly for the social rate of return, but certainly not for the private rate of return. What the graph suggests is a more tightly packed distribution, *except* for the outlier.

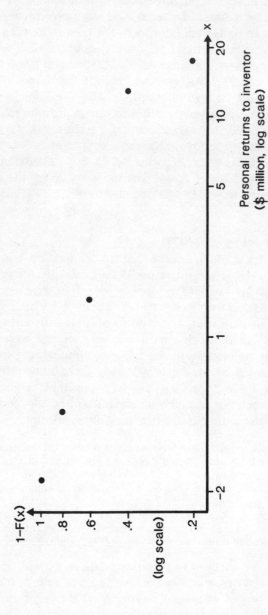

Figure 8.6: Profitability of successful inventions in petroleum refining

Note: Only five observations displayed (excludes salaried inventors). The slope α appears to lie between 0.25 and 0.35.
Source: J. L. Enos (1962).

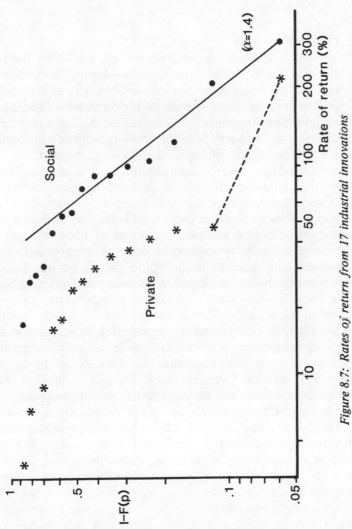

Figure 8.7: Rates of return from 17 industrial innovations

Source: E. Mansfield *et al.* (1977).

Speculating freely, it might be supposed that most of these innovations are routine, run-of-the-mill improvements as indeed the
authors claim, but a few may in fact be of a more radical, 'state
of the art' variety.

(iv) Patent Renewals

In most studies of individual inventions there is a strong 'success
bias'. Since it is difficult, it not by definition impossible, to observe
the output of inventive effort that did not result in an invention;
and since in practice most research has concentrated on the 'noteworthy' inventions that have thrust themselves on the researchers,
what tends to be measured is in the upper tail of the distribution
of inventions by value. When attempting to examine the supply
response of inventions to inducements on the demand side,
however, it is important to gather evidence on marginal inventions
and, if possible, infra-marginal inventions.

As Sanders' data seem to suggest, most inventions are just about
marginal one way or the other. That is, the distribution seems to
be clustered around the breakeven point of profitability. This
impression is corroborated by the figures for lapsed patents and
patent renewals in the UK. These UK data are interesting because
a tax is levied on the patent monopoly according to the length of
duration of the patent. About one-fifth of patents survive for the
maximum duration of 16 years, and these are presumably in the
main protecting economically valuable inventions. For the remainder,
the survival of a patent for another year depends on the patent-
holder's willingness to pay the renewal fee, and this in turn must
reflect his expectation as to the invention's value. In addition it reflects a process of learning as the patent holder becomes more aware
of the invention's qualities as it is being developed and appraised.

Data on UK patent renewals are pictured in Figure 8.8. Note that
this figure has arithmetic as opposed to logarithmic scales for the
axes. It gives an impression of a rectangular frequency distribution
for infra-marginal patents, but this may be a little misleading since
the renewal fee increases with the age of the patent. In any event
this distribution for run-of-the-mill patents is quite unlike the
foregoing cases which examined expressly successful inventions.

To summarise this section, if the *ex-ante* distribution of profitability
of inventions mirrors that of the *ex-post* actual distribution, then

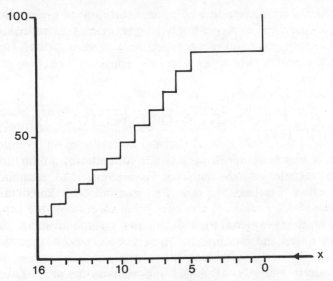

Figure 8.8: Cumulative percentage of UK patents renewed for x years

Note: Shows patents granted in period 1954–58; initially 105,500.
Source: C. Taylor and Z. A. Silbertson (1973).

at low supply prices the supply function will be rather inelastic. The elasticity is given by the Pareto α exponent, which for successful inventions might be around one half. This is the evidence, such as it is, from the upper tail of the distribution of inventions by net value. However, the numerical bulk of the distribution is concentrated around the margin of profitability, and here the elasticity of the cumulative distribution function is quite high. Hence for 'run-of-the-mill' inventions the supply response measured in terms of numbers of inventions is elastic. This is on the assumption that there is a decent correlation between *ex-ante* and *ex-post* profitability so that inventors can screen their research projects. But it was suggested by Mansfield that expected private profitability is not well related to *ex-post* social profitability. If this lack of correlation carries over to *ex-ante* and *ex-post* private profitabilities, it would follow that the supply elasticity of inventions should be highly elastic at a price equal to that corresponding to an average rate of return for all invention projects. This would be a consequence of the inability of inventors to screen their research projects by profitability. It can be seen that this reinforces the idea that there may be a highly elastic response of minor inventions to an increase in their demand price. Accordingly it appears useful to distinguish between important and minor inventions, as the responsiveness of supply to profitability appears to differ considerably between these categories.

8.4   THE DISTRIBUTION OF INVENTORS BY PRODUCTIVITY

Another way to approach the question of diminishing returns, and hence inelasticity in the supply of inventions, is to examine the production function directly i.e. examine the input–output relation.

As Machlup suggests, there can be two variations on this theme – one micro and one macro. The micro evidence relates to the productivity of the individuals in the invention labour force, while the macro evidence examines the relation between research resources and research output in the aggregate. The latter approach is treated in section 8.5.

The idea behind the micro evidence on the input–output relation for inventive activity is rather similar to that presented in the previous section on the distribiution of inventions by value. It is to examine the distribution of inventors by productivity.

Again, the evidence is not exactly plentiful, but one must make do with what one can get. What data there are measure the output of research workers by a count of inventions or research papers, but do not value the items of that count. It is therefore assumed that the average value of an invention from a prolific inventor or research team is the same as that from a typical one of the numerous one-off inventors. The productivity or value of an inventor or inventive team is then measured by how many inventions he or they produce.

Evidence on researcher-output in government and business laboratories was presented in the mid-1950s by William Shockley (1957)[7] (see Figures 8.9 and 8.10). It appears that the cumulative distribution of research workers by productivity, as measured by the number of patents or research publications, is roughly linear in the logarithm of productivity. This suggests an exponential distribution, though the upper tail is more skewed here.

For example, in one large industrial laboratory employing about

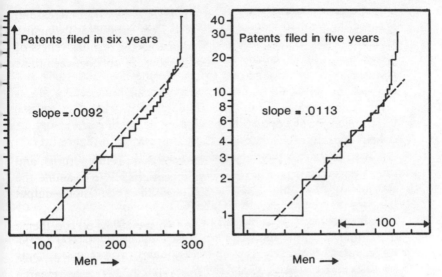

Figure 8.9: The distribution of research workers by productivity
Note: Cumulative distribution on logarithmic scale for patents at two large industrial laboratories.
Source: Shockley (1957).

280 researchers, roughly 100 had filed no patents in a six-year period, about 30 filed 1 patent, while the most productive worker filed approximately 100 patents. The same pattern was repeated in the four other laboratories catalogued by Shockley.

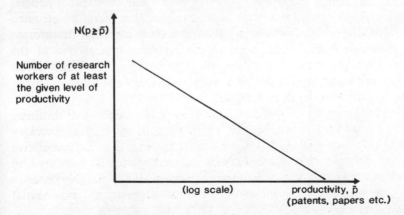

Figure 8.10: Schematic presentation of implicit marginal product schedule

If possible interaction effects between research workers are ignored, figure 8.10 may be interpreted as a marginal productivity schedule. For this case, in which the productivity of any individual researcher is independent of the presence or absence of other researchers, it is only needed to reverse the horizontal axis and convert the log scale of productivity into an arithmetic scale in order to get the derived demand for research workers. Of course, it is implicit for this operation that the unit value of research output is held constant. This interpretation is presented in figure 8.11.

Since the curve is based on observation, it is presumably the case that the researchers whose productivities were measured are all supra-marginal. So, to interpret the curve as derived demand requires the vertical axis to be labelled as the positive deviation of researcher salaries from the current norm. With this adjustment the figure indicates how many researchers would be employed at higher wage rates. However, the figure is mainly impressionistic. It suggests high elasticity of demand for researcher services at current costs of such services, with the elasticity diminishing sharply as costs increase.

Another interpretation of the schematic evidence displayed in Figure 8.9 is with regard to the marginal cost of research output, i.e. the supply schedule for research output. Assuming that the laboratories minimised cost for the particular research output they achieved, then the marginal cost of an invention is the cost of the marginal researcher divided by his output. The margin in question is the 'extensive margin', to use Ricardian terminology. That is,

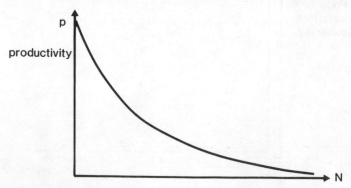

Figure 8.11: Derived demand (marginal product) for researchers

it does not take into account variations in *individual* researcher productivity that are affected by incentives.[8] This does not seem unreasonable for the data provided by Shockley, since variations in researcher salaries were much less spread out than those of productivity, and were only weakly correlated with it.

The picture that emerges in this case would look something like that shown in Figure 8.12, where the marginal cost or supply curve is much more inelastic at the current margin of output (i.e. at the current observed price or value of inventions) than it would be at lower levels of output.

The contention here relating to increasing marginal cost is based on the observed wide disparities in the productivity of research workers and is very much in the spirit of Machlup's 'second shrinkage', which he attributed to lower quality of the personnel. Machlup is implicitly considering the effects of an expansion of inventive activity when he says: 'a point must exist beyond which further transfers to the research and development work force cannot possibly be of the same quality.'

In this context of inventor productivity it is interesting to note some observations in the parallel field of researcher productivity in science. Taking the basic measure of scientific output to be a count of publications, a remarkable regularity was discovered in the 1920s by the biomathematician Lotka (1926). Lotka's law relates the number of authors $A(n)$ who publish exactly n papers by an

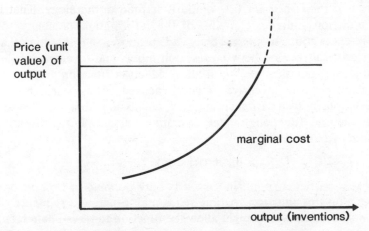

Figure 8.12: Marginal cost of inventive output

inverse square proportionality:

$$A(n) \sim 1/n^2$$

Lotka's law relates to whole disciplines of science. However, it resembles in character the laboratory-worker productivity figures given by Shockley, though it depicts an even more skewed distribution.[9]

In summary, it can be seen that, in contradistinction to the implications in the previous section on the distribution by value of invention projects, the distribution by productivity of research workers is likely to be an important source of inelasticity of supply of inventions at their current price, with the elasticity increasing rather than decreasing as that price is lowered. Since, as Machlup pointed out, the overall effect of these independent sources of supply inelasticity is cumulative, it follows that, taken together with the previous section, the supply of inventions may indeed be rather inelastic over a wide range of unit values of inventive output.

8.5   THE 'INVENTION PRODUCTION FUNCTION'

Supply functions are in principle derived under an assumption of cost-minimisation subject to a production function constraint. They represent the relation between the auxiliary variable (Lagrange multiplier), which is the marginal cost of output, and level of the constraint (i.e. output). It follows therefore that the production function contains all the information necessary for inferences about supply elasticity. It is proposed in this section that a statistical relation between research inputs and inventive output may trace out such an 'invention production function'.

Consider the inventive activity engaged in as research and development by a particular industry, and assume that an invention production function relates inventive output I to inventive inputs R:

$$I = f(R), \quad f'(R) > 0, \quad f''(R) < 0$$

At a particular point in time it is only possible to observe one point on this function. A time-series may generate a sequence of distant points which might allow the function to be estimated, but this requires (a) an assumption that f( ) is stable over time, and (b) in practice more variability in R than might reasonably be expected. On the other hand, to estimate the function f( ) from cross-section inter-industry data would require an implausible assumption that the function is the same for all industries. Since 'technological opportunity', as Scherer (1965) calls it, obviously does vary between industries it would appear that n observations on n industries offers no improvement over one observation from one industry.

It may be possible, however, to replicate an observation at a particular time for a particular industry by considering variation across countries. At least the knowledge base or state of the art for a technology, which conditions the function f( ), is to a first approximation common across countries. This therefore offers some scope for an empirical estimation of the relation between inventive inputs and output.

A principal cause of diminishing returns in the function f( ) is, as Machlup pointed out, the non-homegeneity of the research labour force, which was examined in some detail in the previous section. If the research labour force were quite homogeneous it may be expected that doubling the number of researchers would result in twice the inventive output. It would also be expected that the same volume of research labour inputs in different countries would lead to equal inventive output. But 100 research workers in a small country might represent 10 per cent of the research labour force whereas the same number of researchers might be only 1 per cent of the research labour force in a large country. Obviously the best 100 researchers from a country with a large research pool is likely to be more productive than the best 100 from a country with more limited research resources. Hence the function f( ) should differ between countries.

How the invention production function differs between industries and countries must now be considered. The simplest assumption to make regarding the differential effects of technological opportunity between industries would be to assume that, for given inputs, the inventive outputs of two industries always stand in the same proportion. Similarly for the inter-country differences it might be assumed that for given research inputs the inventive outputs would again stand in a constant proportion. Moreover it would be reasonable to suppose that these sources of production function shift are independent.

If the production function is now specialised to a constant elasticity form:

$$f(R) = A.R^{\alpha}$$

the foregoing assumption may be represented as:

$$I_{ij} = A_{ij}.R_{ij}^{\alpha}$$

or $\ln(I_{ij}) = \ln(A_{ij}) + \alpha.\ln(R_{ij})$

in which I stands for invention, subscript i represents 'industry' and subscript j represents 'country'. Of course, if it is believed that technological opportunity might differentially affect the elasticity between industries, then the parameter α too should be subscripted with i.

As in the preceding chapters, a count of inventions is used as the measure of inventive output. In particular in this section inventive output is measured by a count of international patents. These are patents that are taken out simultaneously in more than one country. Because the cost of obtaining an international patent is considerably larger than that of obtaining a patent in only one country, it follows that they typically represent more important inventions.

It is true that more direct measures of technical progress than patents are conceivable, such as the rate of advance of total factor productivity or decline in unit production costs or diversification of products, but these all suffer a disadvantage compared to patents in that technology is international and so these measures of progress may in fact have been produced in large part elsewhere. A virtue of patents is that they are unambiguously related to the inventive activity that produced the invention.

What is reported below therefore is a study of the association between R & D expenditures (input) and counts of international patents (output) across 15 industries and 7 countries. The patents were counted as applications made in 1975 for a wide, but not exhaustive, selection of technology classes (see US Department of Commerce, 1977; and the R & D expenditures are those reported by the OECD, 1976 for 1972. The R & D expenditures are denominated in US $; no attempt was made to adjust the exchange rates for differing real costs of scientific and technical manpower across countries. Although the R & D data are classified by industry, the international patent data were not classified by industries in a directly comparable manner, so it was necessary to use an element of judgement to obtain a concordance between the quasi-technological patent classification and the industrial classification of expenditures. Thus the industry categories used in the analysis of covariance reported below allow for differences not only in technological opportunity but also in the propensity to patent and, hopefully, in the matching of the patent and expenditure statistics. In a similar vein, the country categorisation not only reflects a production function shift but also differing propensities to take out an international patent as between countries.

The scatter plot shown as Figure 8.13 indicates that:

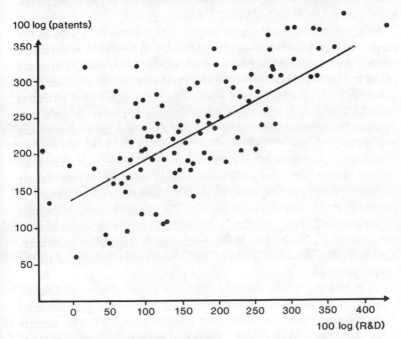

Figure 8.13: International patents against R and D expenditure

Regression equation 1: $\ln(I_{it}) = A + \alpha.\ln(R_{it})$
$$= 3.54 + 0.516.\ln(R_{it})$$
$$(0.27) \ (0.060)$$
$$r^2 = 0.444, \quad \text{S.E.E.} = 1.322$$

Regression equation 2: $\ln(I_{it}) = A_i + A_j + \alpha.\ln(R_{it})$
$$= \{\text{ind. \& country effects}\} + 0.175.\ln R_{it}$$
$$(0.045)$$
$$R^2 = 0.972, \quad \text{S.E.E.} = 0.336$$

(i)   Plotting logarithm of patent counts against the logarithm of
      R & D expenditures ($ million) reveals an upward drift with a
      lot of variability around it;
(ii) The spread of observations around the regression line tends to
      increase at lower levels of R & D. This is a slightly unusual
      form of heteroskedasticity, but one that can be explained by
      supposing that an additive constant is present in the underlying
      relationship. If this were the case it could be described as the
      autonomous component of invention; and
(iii) The curve is linear in log-logs with a slope of about a half, sug-
      gesting the application of a square root law for the production
      function.[10] It implies an elasticity of inventive output with
      respect to R & D inputs of a half.

The first regression simply gives the line shown in Figure 8.13.
Regression equation 2, which takes additive industry and country
effects into account simultaneously with an estimate of the slope or
elasticity, yields a rather low elasticity. The difference between this
estimate of 0.175 and the elasticity of about one half from the
simple regression is accounted for statistically by the fact that the
industry and country effects are correlated with the explanatory
variable $\ln(R_{it})$. Further regressions were calculated, including one
which allowed the slope coefficients to differ across industries – this
restored a typical elasticity of around one half, but none of the
individual industry slope coefficients proved statistically significant.
It might have been possible to explore the suggestion above about
the source of the heteroskedasticity by using nonlinear regression
methods, but the quality of the data did not warrant such extensive
experimentation.

It seems therefore that the elasticity of inventive output with
respect to research and development expenditures probably lies in
the range 0.1 to 0.5. Now with constant elasticity invention production function $I = A.R^{\alpha}$, the marginal productivity of research is
$\alpha A R^{\alpha-1}$, from which the marginal cost of invention may be derived
as the inverse since the cost of an increment in R when R is
measured by expenditure is definitionally unity. It follows therefore
that the elasticity of supply is $\alpha/(1 - \alpha)$, which is therefore in the
range 0.1 to 1.0. In addition if it were true that the observed
heteroskedasticity could be due to a 'threshold effect' as tentatively
suggested above, that would make the supply function more
inelastic.

8.6 SUMMARY AND CONLUSIONS

This chapter complements the preceding two chapters in attempting to assess the view that invention is an endogenous variable in economics. As in Chapter 7, the question of endogeneity is assessed in the context of a supply function. However, in this chapter the supply function is approached indirectly. Rather than directly relating the volume of inventive activity to the price signal that induces it, the sources of inelasticity are examined. The essential idea is that the most profitable inventive projects and the most productive researchers will be called on first in a kind of filtering process. This view highlights the distribution of projects by profitability and the distribution of inventors by productiveness. Both distributions are very spread out and very left-skewed, implying in the argument of this chapter an inelastic supply. Apparent confirmation of this inelasticity of supply is presented in section 8.5 which attempts a production function interpretation of international inter-industry data on patents and research and development expenditure.

NOTES

1. Whether this supposition is correct is not know – on the whole one supposes that people are risk-averse, but this may not be true for inventors as a class. They may be 'plungers' rather than 'hedgers'.
2. An important literature has been developed in recent years in the economics of screening, particulary with respect to the labour market. See, for example, Stiglitz (1975).
3. In surveys of research and development the activity is often divided into the trinity: basic research; applied research; experimental development. The OECD's *Frascati Manual* (1975), which is a proposal for standardising such surveys, describes applied research as:

   undertaken either to determine possible users for the findings of basic research or to determine new methods or ways of achieving some specific and pre-determined objectives. It involves the consideration of the available knowledge and its extension in order to solve particular problems. In the Business Enterprise sector the distinction between basic and applied research will often be marked by the creation of a new project to explore any promising results of a basic research programme . . . . Applied research develops ideas into operational form.

   While this description, and the examples given in the same publication to distinguish the three categories, makes clear the fact that applied research as measured in the surveys is wider than a screening process of potential invention research projects for the investment of resources at the development stage, it is also clear that this could be an important part of such activity.

4. For example, see K. Hartley and J. Cubitt (1976/77)
5. If the variable, u, is distributed with a distribution function (cumulative frequency) $F(u)$, then the Pareto distribution holds if $\log(1 - F(u))$ is linearly related to $\log(u)$. Mandelbrot (1963) has pointed out that a visual test of approximate linearity is valid if the line has a shallow slope, which is a measure of the exponent, α, in the distribution function.

   One of the features of the Pareto distribution that makes it uncomfortable to analyse or use in statistical methodology is the fact that for $\alpha < 1$ it possesses neither mean nor variance and for $\alpha < 2$ the variance does not exist.
6. Nordhaus (1969) noted Scherer's observation that the exponent of the Pareto distribution was 'equal to' (in fact Scherer says 'less than') 0.5 and that this further implies that it is not possible to reduce risk by carrying a suitably diversified portfolio of research projects. He also noted that Machlup had come to a similar conclusion by an unspecified route. Machlup says:

   Contrary to other industries where the probable errors are larger for individual producers than for industry as a whole, the 'invention industry' is apt to present smaller dispersions in the probability distributions for the individual producers.

7. William Shockley, who invented the transistor, was awarded the Nobel prize in physics for his invention.
8. Thomas Edison's famous quotation that 'Genius is one per cent inspiration and ninety-nine per cent perspiration' could be interpreted as a statement about the intensive margin of productivity for researchers who are extremely supramarginal on the extensive margin. Apart from the chance element in inspiration it suggests a rough proportionality between inventive output and inputs of time and effort. Arthur Koestler (1964) notes that 'At a time when his inventions were transforming the pattern of our civilisation, "his [Edison's] ignorance of scientific theory raised criticism and opposition, especially among highly trained scientists and engineers without inventive talent"' (quotation from J. G. Crowther, 1947). This again suggests the importance of the inter-individual extensive margin of productivity, which since it is largely innate, must be treated as an exogenous datum rather than an endogenous variable in economic analysis.
9. Nicholas Rescher, who draws attention to Lotka's law in his book entitled *Scientific Progress*, makes the tantalising suggestion that it may be possible to infer something about the distribution of the *quality* of scientific results from the development of their *quantity*. In

particular, he argues that while there has been an exponential growth in the number of total scientific findings, the number of first-rate findings has been growing approximately linearly. He says that, with the total number of findings Q, the volume of 'λ-quality findings stands at Q^λ (for $0 < \lambda < 1$)'. And he gives labels for the various values of λ:

$\lambda = 1$: routine
$\lambda = 1/2$: important ('Rousseau's law')
$\lambda = 1/4$: very important
($\lambda = 0$), log Q: first rate.

If it were possible to measure scientific quality, then on this basis we should observe a double-logarithmic rank-size correlation (Zipf's law). The evidence of the distribution by value of inventions suggesting a possibly Pareto distribution would be consistent with this suggestion of a quality–quantity relationship, if one interprets importance as profitability.

10. If this is interpreted according to Rescher's schema outlined in note 9 above, it suggests that internationally patented inventions represent R & D output of 'important' quality level, while the national income accounting convention of measuring output by inputs (as is done for all service sectors including R & D) is implicitly measuring 'routine' inventive output as a benchmark.

References

Abramovitz, M. (1956) 'Resource and output trends in the United States since 1870', *American Economic Review*, 46 (2), May, 5–23.

Ahmad, S. (1966) 'On the theory of induced invention', *Economic Journal*, 76 (2), June, 344–57.

Arrow, K. J. (1962) 'Economic welfare and the allocation of resources for invention', in R. R. Nelson (ed.), *The Rate and Direction of Inventive Activity*, NBER, Princeton University Press, Princeton, N.J.

Barzel, Y. (1968) 'Optimal timing of innovations', *Review of Economics and Statistics*, 50 (3), August, 348–55.

Baumol, W. J. (1962) *Business Behavior, Value and Growth*, Harcourt, Brace, Jovanovich Inc., New York.

Baumol, W.J., Panzar, J. C. and Willig, R. D. (1982) *Contestable Markets and the Theory of Industry Structure*, Harcourt, Brace Jovanovich, San Diego, Ca.

Baumol, W. J. and Willig, R. D. (1981) 'Fixed costs, sunk costs, entry barriers and the sustainability of monopoly', *Quarterly Journal of Economics* 95 (3) August 405–31.

Bergstrom, V. and Melander, M. (1979): 'Production functions and factor demand functions in postwar Swedish industry', *Scandinavian Journal of Economics*, 81 (4), 534–51.

Berndt, E. R. and Khaled, M. S. (1979) 'Parametric productivity measurement and choice among flexible functional forms', *Journal of Political Economy*, 87 (6), December, 1220–45.

Bernouilli, J. (1713) *Ars Conjectandi*, Basle.

Bhagwati, J. (1982) 'Directly unproductive, profit-seeking (D.U.P.) activities', *Journal of Political Economy*, 90(5), October, 988–1002.

Binswanger, H. P. (1974) 'A microeconomic approach to induced innovation', *Economic Journal* 84 (336), December, 940–58.

Binswanger, H. P. (1974) 'The measurement of technical change biases with many factors of production', *American Economic Review* 64 (5), December, 964–76.

Binswanger, H. P. and Ruttan, V. W. (eds) (1978) *Induced Innovation*, Johns Hopkins University Press, Baltimore.

Box, G. E. P. and Draper, N. R. (1969) *Evolutionary Operation*, Wiley, London.

Buchanan, J. M., Tullock, G. and Tollison, R. D. (1980) *Towards a Theory of the Rent-Seeking Society*, Texas A. & M. Press.

Burmeister, E. and Dobell, A. R. (1969) 'Disembodied technical change with several factors', *Journal of Economic Theory* 1 (1), June, 1–8.

Burmeister, E. and A. R. Dobell (1970) *Mathematical Theories of Economic Growth*, Macmillan, London.

Burns, A. (1934) *Production Trends in the United States since 1870*, National Bureau of Economic Research.

Cox, R. T. (1961) *The Algebra of Probable Inference*, Johns Hopkins University Press, Baltimore.

Crowther, J. G. (1947) *British Scientists of the Nineteenth Century*, Penguin, London.

Dasgupta, P. and Stiglitz, J. E. (1980) 'Industrial structure and the nature of innovative activity', *Economic Journal* (2), June, 266–93.

Dasgupta, P. and Stiglitz, J. E. (1980) 'Uncertainty, industrial structure and the speed of R & D', *Bell Journal of Economics* 11 (1), Spring, 1–28.

David, P. A. and van de Klundert, T. (1965) 'Biased efficiency growth and capital–labor substitution in the US, 1899–1960', *American Economic Review*, 55 (3), June, 357–94.

Demsetz, H. (1969) 'Information and efficiency: another viewpoint', *Journal of Law and Economics*, 12, 1–22.

Descartes, R. (1967) *Discourse on Method*, Everyman edition, Dent, London.

Drandakis, E. M. and Phelps, E. S. (1966) 'A model of induced inventions, growth and distribution', *Economic Journal*, 76 (304), December, 823–40.

Enos, J. L. (1962) 'Invention and innovation in the petroleum refining industry', in R. R. Nelson (ed.), *The Rate and Direction of Inventive Activity*, NBER, Princeton University Press, Princeton, N.J.

Frisch, R. (1933) 'Propagation problems and impulse problems in dynamic economics', in *Essays in Honour of Gustav Cassel*, 171–206, Allen and Unwin, London.

Geweke, J., Meese, R. and Dent, W. T. (1983) 'Comparing alternative tests of causality in temporal systems: analytic results and experimental evidence', *Journal of Econometrics*, 21 (2), February 161–94.

Gordon, H. S. (1954) 'The economic theory of a common-property resource: the fishery', *Journal of Political Economy*, 62 (2), March, 124–42.

Granger, C. W. J. (1969) 'Investigating causal relations by econometric models and cross-spectral methods', *Econometrica* 37 (3), July, 424–38.

Griliches, Z. (1957) 'Hybrid corn: an exploration in the economics of technical change', *Econometrica* 25 (4), October, 501–22.

Griliches, Z. and Schmookler, J. (1963): 'Inventing and maximizing', *American Economic Review*, 53 (4), September, 725–9.

Guilkey, D. K. and Salemi, M. K. (1982) 'Small sample properties of three tests for Granger-causal ordering in a bivariate stochastic system', *Review of Economics and Statistics*, 65, 4 (Nov.), 668–80.

Hadley, G. (1968) *Nonlinear and Dynamic Programming*, Addison-Wesley, Reading, Mass.

Hartley, K. and Corcoran, W. (1978) 'The time-cost trade-off for airliners', *Journal of Industrial Economics* 24 (3), 209–22.

Hartley, K. and Cubitt, J. (1976/77) 'Cost escalation in the UK', *Civil Service Expenditure Committee*, Appendix 44, House of Commons Papers 535-III HMSO.

Harvey, A. C. (1981) *The Econometric Analysis of Time Series*, Philip Allan, Oxford.

Haugh, L. D. and Box, G. E. P. (1977) 'Identification of dynamic regression (distributed lag) models connecting two time series', *Journal of the American Statistical Association* 72 (1), 121–30.

Hayami, Y. and Ruttan, V. W. (1971) *Argricultural Development*: an International Perspective, The Johns Hopkins University Press, Baltimore.

Hicks, J. R. (1932) *The Theory of Wages*, Macmillan, London (2nd edn, 1964).

Hu, S. C. (1973) 'On the incentive to invent: a clarificatory note', *Journal of Law and Economics*, 16, 169–78.

Jeffreys, H. (1961) *Theory of Probability*, Oxford University Press, Oxford.

Jewkes, J., Sawers, D. and Stillerman, R. (1958; 2nd edn, 1969) *The Sources of Invention*, Macmillan, London.

Johansen, L. (1959) 'Substitution versus fixed coefficients in the theory of economic growth', *Econometrica*, 27(2), April, 157–76.

Jorgenson, D. W. and Fraumeni, B. (1981) 'Relative prices and technical change' in E. R. Berndt and B. S. Field (eds), *Modelling and Measuring Natural Resource Substitution*, M.I.T. Press, Cambridge, Mass.

Kamien, M. and Schwartz, N. (1969) 'Induced factor augmenting technical change from a microeconomic viewpoint', *Econmetrica* 37 (4), October, 668–84.

Kamien, M. I. and Schwartz, N. L. (1982) *Market Structure and Innovation* Cambridge University Press, Cambridge.

Kay, N. M. (1979) *The Innovating Firm*, Macmillan, London.

Kennedy, C. (1964) 'Induced bias in innovation and the theory of distribution', *Economic Journal*, 74 (3), September, 541–7.

Keynes, J. M. (1921) *A Treatise on Probability*, Macmillan, London.

Khinchine, A. I. (1957) *Mathematical Foundations of Information Theory*, Dover, London.

Kitch, E. W.(1977) 'The nature and function of the patent system', *Journal of Law and Economics* 20 (2) October, 265–90.

Koestler, A. (1964) *The Act of Creation*, Hutchinson, London.

Krueger, A. O. (1974) 'The political economy of the rent-seeking society', *American Economic Review*, 64 (3), June, 291–303.

Kuznets, S. (1936) *Secular Movements in Production and Prices* Houghton Mifflin, New York.

Lancaster, K. (1966) 'A new approach to consumer theory', *Journal of Political Economy*, 74 (2), April 132–57.

Lee, T. and Wilde, L. L. (1980) 'Market structure and innovation: a reformulation', *Quarterly Journal of Economics*, 94 (1), February 429–36.

Leibenstein, H. (1966) 'Allocative efficiency vs. "X-efficiency"', *American Economic Review* 56 (2), June, 392–415.

Lotka A (1926): 'The frequency distribution of scientific productivity', *Journal of the Washington Academy of Science*, 16.

Loury, G. C. (1979) 'Market structure and innovation', *Quarterly Journal of Economics*, 93 (3), August, 395–410.

Lynk, E. L. (1982) 'Factor demand, substitution and biased technical change in Indian manufacturing industries', *The Manchester School*, 50 (2), June, 126–38.

Machlup, F. (1962) 'The supply of inventors and inventions', in R. R. Nelson (ed.), *The Rate and Direction of Inventive Activity*, NBER, Princeton University Press, Princeton, N.J.

Mandelbrot, B. (1963) 'New methods in statistical economics', *Journal of Political Economy*, 71 (5), October, 421–40.

Mansfield, E. (1968) *Industrial Research and Technological Innovation*, Longmans, London.

Mansfield, E., Rapoport, J., Schnee, J., Wagner, S., and Hamburger, M. (1977) *The Production and Application of New Industrial Technology*, Norton, New York.

Marris, R. (1964) *The Economic Theory of Managerial Capitalism*, Macmillan, London.

Marshak, T., Glennan, T. K., and Summers, R. (1967) *Strategy for*

R & D: Studies in the Microeconomics of Development, Springer Verlag, Berlin.

Marshall, A. (1890) *Principles of Economics*, (1925 edn), Macmillan, London.

Marshall, A. W. and Meckling W. (1962) 'Predictability of the costs, time and success of development', in Nelson, R. R. (ed.), *The Rate and Direction of Inventive Activity*, Princeton, N.J.

McGee J. S. (1966) 'Patent exploitation: some economic and legal problems', *Journal of Law and Economics*, 9, 135–62.

Merton, R. (1935) 'Fluctuations in the rate of industrial invention', *Quarterly Journal of Economics*, May.

Minasian, J. (1962) 'The economics of research and development', in R. R. Nelson (ed.), *The Rate and Direction of Inventive Activity*, N.B.E.R., Princeton University Press, Princeton, N.J.

Minasian, J. (1969) 'Research and development, production functions and rates of return', *American Economic Review*, 59, May supplement.

Moroney, J. R. and Trapani, J. M. (1981) 'Alternative models of substitution and technical change in natural resource intensive industries', in Berndt, E. R. and Field B. C. (eds), *Modelling and Measuring Natural Resource Substitution*, MIT Press, London.

Mosteller, F. and Rourke, R. E. K. (1973) *Sturdy Statistics*, Addison-Wesley, London.

Mosteller, F. and Tukey, J. W. (1979) *Data Analysis and Regression* Addison-Wesley, Reading, Mass.

National Science Foundation (1972) *Research and Development in Industry 1970*, Surveys of Science Resources Series, Washington, D.C.

Nelson, R. R. (1961) 'Uncertainty, learning and the economics of parallel research and development efforts', *Review of Economics and Statistics* 43 (4), November, 351–64.

Nelson, R. R., Winter, S. G. and Schuette, H. L. (1976) 'Technical change in an evolutionary economy', *Quarterly Journal of Economics* 79 (358), February, 90–118.

Nicoll, G. R. (1981) *The Entropy of a Number of Common Probability Distributions*, Heriot-Watt University, Department of Electrical and Electronic Engineering, Occasional Notes 1981 (2).

Nordhaus, W. D. (1969a) 'An economic theory of technological change', *American Economic Review*, 59 (2) May, 18–28.

Nordhaus, W. D. (1969b) *Invention, Growth and Welfare*, MIT Press, London.

Nordhaus, W. D. (1973) 'Some skeptical thoughts on the theory of induced innovation', *Quarterly Journal of Economics*, 87 (2), May, 208–19.

OECD (1968) *Gaps in Technology between Member Countries*, Organisation for Economic Cooperation and Development, Paris.

OECD (1975) *'Frascati Manual'*: *The Measurement of Scientific and Technical Activities*: *Proposed Standard Practice for Surveys of Research and Experimental Development*, Organisation for Economic Cooperation and Development, Paris.

OECD (1976) *International Statistical Year, 1973*, Organisation for Economic Cooperation and Development, Paris.

Panzar, J. C. and Willig, R. D. (1977) 'Free entry and sustainability of natural monopoly', *Bell Journal of Economics* 8 (1), Spring, 1–22.

Peck, M. and Scherer, F. M. (1962) *The Weapons Acquisition Process*, Harvard University Press, Cambridge, Mass.

Pierce, D. A. (1977) 'Relationships – and the lack thereof – between economic time-series, with special reference to money and interest rates', *Journal of the American Statistical Association* 72 (1), 11–22.

Pierce D. A. and Haugh L. D. (1977) 'Causality in temporal systems', *Journal of Econometrics*, 5 (2) 265–93.

Plant, A. (1934) 'The economic theory concerning patents on inventions', *Economica*, 1(n.s.), February, 30–51.

Posner, R. A. (1975) 'The social costs of monopoly and regulation', *Journal of Political Economy*, 83(4), August, 807–28.

Ramsay, F. P. (1931) 'Truth and probability', reprinted in Kyberg and Smokler (eds) *Studies in Subjective Probability* (1964), Wiley, London.

Rescher, N. (1978) *Scientific Progress*: *a Philosophical Essay on the Economics of Research in Natural Science*, University of Pittsburg Press, Pittsburg, Pa.

Rosenberg, N. (1974) 'Science, invention and economic growth', *Economic Journal*, 84 (1), 90–108 March. Also in Rosenberg N. (1976), Chapter 15.

Rosenberg, N. (1976) *Perspectives on Technology*, Cambridge University Press, Cambridge.

Rosenberg, N. (1982) *Inside the Black Box*, Cambridge University Press, Cambridge.

Salter, W. E. G. (1960) *Productivity nd Technical Change*, Cambridge University Press, Cambridge.

Samuelson, P. A. (1962) 'Parable and realism in capital theory: the surrogate production function', *Review of Economic Studies*, 39 (2), June, 193–206.

Samuelson, P. A. (1965) 'A theory of induced innovation along Kennedy–Weizsacker lines', *Review of Economics and Statistics*, 47 (4), November, 343–56.

Sanders B. S., Rossman, J. and Harris, L. J. (1968) 'The economic impact of patents', in *Patent, Trademark and Copyright Journal of Research and Education*, 2 (3), September, 340–62.

Scherer, F. M. (1965) 'Firm size, market structure, opportunity and the output of patented inventions', *American Economic Review*, 55 (5), December, 1097–125.

Scherer, F. M. (1965) 'Government research and development programs', in R. Dorfman (ed.), *Measuring Benefits of Government Expenditure*, The Brookings Institution, Washington, D.C.

Scherer, F. M. (1966) 'Time–cost tradeoffs in uncertain empirical projects', *Naval Research Logistics Quarterly* 13, 71–82 and 349–50.

Scherer, F. M. (1972) 'Nordhaus' theory of optimal patent life, a geometric reinterpretation', *American Economic Review* 62 (3), June, 422–7.

Scherer, F. M. (1980) *Industrial Market Structure and Economic Performance* (2nd edn.), Rand McNally, Chicago.

Schmitt, S. A. (1969) *Measuring Uncertainty: an Elementary Introduction to Bayesian Statistics*, Addison-Wesley, London.

Schmookler, J. (1962): 'Changes in industry and the state of knowledge as determinants of inventive activity', in R. R. Nelson (ed.), *The Rate and Direction of Inventive Activity*, Princeton, N.J.

Schmookler, J. (1966) *Invention and Economic Growth*, Harvard University Press, Cambridge, Mass.

Schmookler, J. (1972) *Patents, Invention and Economic Change*, Harvard University Press, Cambridge, Mass.

Schmookler, J. and Brownlee, O. (1962) 'Determinants of inventive activity', *American Economic Review* 52 (2), May, 165–76.

Schumpeter, J. A. (1939; 1964) *Business Cycles*, McGraw-Hill, New York.

Schumpeter, J. A. (1942) *Capitalism, Socialism and Democracy*, Harper & Row, New York.

Shaked, A. and Sutton, J. (1983) 'Natural oligopolies', *Econometrica*, 51 (5), 1469–83.

Shaked, A. and Sutton, J. (1985) *Product Differentiation and Industrial Structure*, Paper presented to the annual conference of the European Association for Research in Industrial Economics, Cambridge, UK.

Shephard, R. W. (1970) *Theory of Cost and Production Functions*, Princeton University Press, Princeton, N.J.

Shockley, W. (1957) 'On the statistics of individual variation of productivity in research laboratories', *Proceedings of the Institute of Radio Engineers*, 45, March, 279–90.

Siegel, I. H. (1962) 'Scientific discovery and the rate of invention' in R. R. Nelson (ed.), *The Rate and Direction of Inventive Activity*, NBER, Princeton, N.J.

Silberberg, E. (1978) *The Structure of Economics: a Mathematical Analysis*, McGraw-Hill, New York.

Simon, H. A. (1955) 'A behavioural model of rational choice', *Quarterly Journal of Economics*, 69, 99–118.

Sims, C. A. (1972) 'Money, income and causality', *American Economic Review*, 62 (4), September, 540–52.

Slutsky, E. (1927) 'The summation of random causes as the source of cyclic processes' (in Russian), *Problems of Economic Conditions* 3, 1. English translation in *Econometrica* 5 (105), 1937.

Smith, A. (1776) *An Inquiry into the Wealth of Nations*, London.

Solow, R. M. (1957) 'Technical change and the aggregate production function', *Review of Economics and Statistics*, 39 (3), August, 312–30.

Solow, R. M. (1958) 'A skeptical note on the constancy of relative shares', *American Economic Review*, 48 (4), September, 618–31.

Stigler, G. (1961) 'The economics of information', *Journal of Political Economy*, 69(3), June, 213–25.

Stiglitz, J. E. (1975) 'The theory of "screening", education and the distribution of income', *American Economic Review*, 65 (3), June, 283–300.

Stoneman, P. (1983) *The Economic Analysis of Technological Change*, Oxford University Press, London.

Swamy, P. A. V. B. (1970) 'Efficient inference in a random coefficient regression model', *Econometrica*, 38 (2), March, 311–23.

Tandon, P. (1982) 'Optimal patents with compulsory licensing', *Journal of Political Economy* 90 (3), June, 470–86.

Taussig, F. W. (1915) *Inventors and Money Makers*, Macmillan, New York.

Taylor, C. T. and Silberston, Z. A. (1973) *Economic Impact of Patents*, Cambridge University Press, Cambridge.

Tobin, J. (1967) 'Comments', in M. Brown (ed.), *The Theory and Empirical Analysis of Production*, NBER Studies in Income and Wealth, vol. 31, 50–3, Columbia University Press, New York.

Tribus, M. (1969) *Rational Descriptions, Decisions and Designs*, Pergamon, London.

United States Department of Commerce (1975) *Historical Statistics of the US, Colonial Times to 1970*, Bureau of the Census, Washington, D.C.

United States Department of Commerce (1977) *Technology Assessment and Forecast*, 7th Report, Office for Technology Assessment and Forecast, Washington D.C.

240 *The Economics of Invention*

Varian, H. (1978) *Microeconomic Analysis*, W. W. Norton & Co. London.

Weitzman, M. (1974) 'Free access vs. private ownership as alternative systems for managing common property', *Journal of Economic Theory* 8 (2), June, 225–34.

Wills, J. (1979) 'Technical change in the US primary metals industry', *Journal of Econometrics* 10 (1), February, 85–98.

Williamson O. E. (1964) *The Economics of Discretionary Behavior*, Prentice-Hall, Englewood Cliffs, N.J.

Williamson, O. E. (1975) *Markets and Hierarchies: Analysis and Antitrust Implications*, Free Press, New York.

Wright, B. D. (1983) 'The economics of invention incentives: patents, prizes and research contracts', *American Economic Review* 73 (4), September, 691–707.

Wyatt, G. J. (1983) *Multifactor Productivity Change in Finnish and Swedish Industries, 1960 to 1980*, ETLA series B no. 38, Helsinki.

Yamey, B. S. (1970) 'Monopoly, competition and the incentive to invent: a comment', *Journal of Law and Economics*, 13, 253–6.

Index of Names

Subject Index